Introducing Meteorology

Companion titles

Introducing Geology – A Guide to the World of Rocks (Second Edition 2010)
Introducing Palaeontology – A Guide to Ancient Life (2010)
Introducing Volcanology – A Guide to Hot Rocks (2011)
Introducing Geomorphology – A Guide to Landforms and Processes (2012)
Introducing Tectonics, Rock Structures and Mountain Belts (2012)
Introducing Oceanography (2012)

For further details of these and other Dunedin Earth and Environmental Sciences titles see:
www.dunedinacademicpress.co.uk

ISBN 978-1-906716-21-9

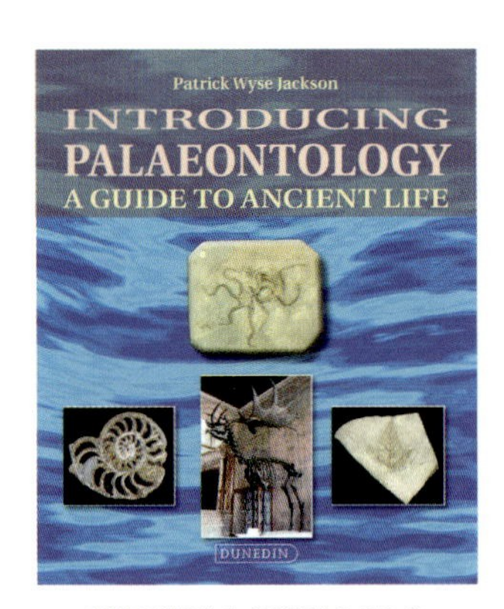

ISBN 978-1-906716-15-8

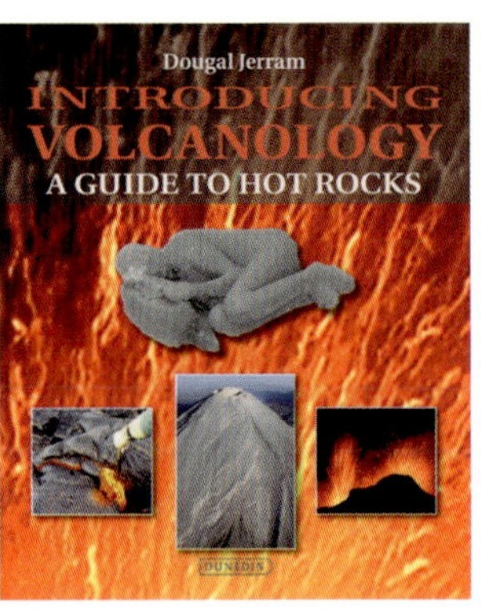

ISBN 978-1-906716-22-6

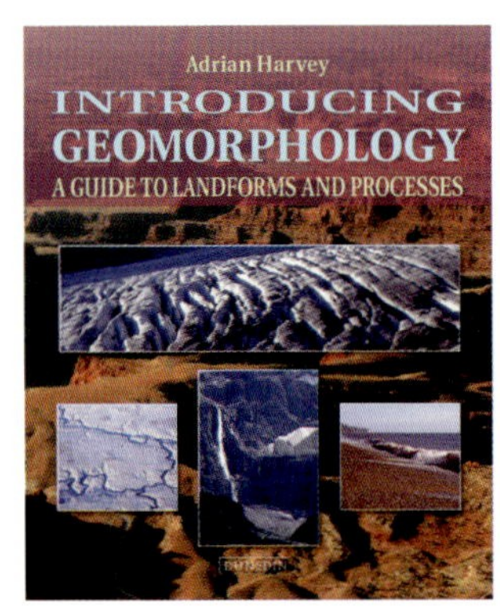

ISBN 978-1906716-32-5

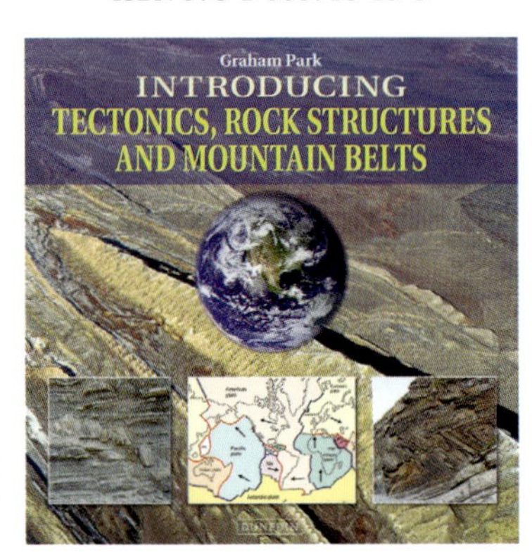

ISBN 978-1-906716-26-4

ISBN 978-1-78046-001-7

Introducing Meteorology

A Guide to Weather

Jon Shonk

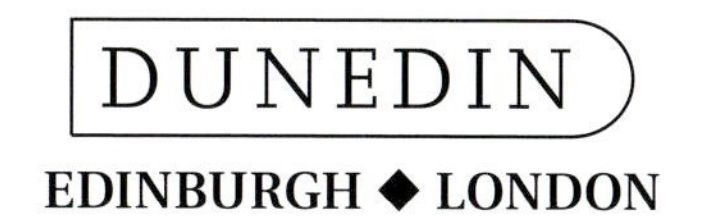

EDINBURGH ◆ LONDON

First published in 2013 by
Dunedin Academic Press Ltd

Head Office: Hudson House,
8 Albany Street, Edinburgh EH1 3QB

London Office: The Towers,
54 Vartry Road, London N15 6PU

Paperback book: 9781780460024
ePub: 9781903544570
ePub (Amazon/Kindle): 9781903544747
ePub (iPad, Fixed Layout): 9781903544754

British Library Cataloguing in Publication data
A catalogue record for this book is available from the British Library

Typeset by Makar Publishing Production, Edinburgh
Printed in Poland by Hussar Books

Contents

Acknowledgements

First, I wish to thank Ross Reynolds and Pete Inness for encouraging me to write this book – their help and support throughout the writing process has been invaluable, as has their willingness to check through various chapters of the manuscript. I also thank Keith Shine for his careful review of Chapter 16, Mike Stroud and Giles Harrison for permitting me to take photos and use data from the Atmospheric Observatory at the University of Reading, and David McLeod and Anne Morton at Dunedin Academic Press for their support and guidance. I must also thank all of the following for agreeing to look through a chapter or two: Lesley Allison, Laura Baker, Andy Barrett, Sylvia Bohnenstengel, Kirsty Hanley, Emma Irvine, Nick Klingaman, Keri Nicoll, Daniel Peake, Ali Rudd, Claire Ryder, Jane Shonk, Peter Shonk, Claire Thompson, Rob Thompson, Andy Turner and Curtis Wood. Finally, I thank all my friends and family for being supportive, and my housemate, Sam Ridout, for constantly reminding me that I should really be writing my book whenever she found me in front of the TV of an evening.

This book is dedicated to the memory of David Grimes.

Figure Credits

Figures 0.1, 11.4, 13.2B, 13.4, 13.8: NASA Earth Observatory.

Figure 1.1: © Rachel O'Byrne.

Figures 1.2, 2.1, 2.2, 3.7ABC, 7.4AB, 7.5, 8.4, 12.1B, 15.5AD: © Jon Shonk.

Figures 1.3, 7.7A: © Michelle Cain.

Figures 2.3ABCD, 3.2AB, 3.3, 3.4, 3.5AB, 3.6AB, 4.1, 4.2, 8.1, 10.9: courtesy of the Department of Meteorology, University of Reading.

Figure 2.4: NOAA Photo Library/US Weather Bureau.

Figures 2.5, 2.6: NOAA Photo Library.

Figure 2.7: US Army Photo.

Figures 3.1, 8.7, 9.7AB, 15.1, 15.2: European Centre for Medium-Range Weather Forecasts (ECMWF).

Figures 3.7DF, 6.5: © Mike Blackburn.

Figure 3.7E: © Peter Smith.

Figure 3.8: NOAA (National Data Buoy Center).

Figures 4.3, 4.4, 10.2, 14.1, 15.3: © Crown Copyright, Met Office.

Figures 4.7ABC, 10.5, 10.11, 12.2, 13.2A: © NEODAAS/University of Dundee.

Figure 5.1: NASA Earth Observatory/JSC Gateway to Astronaut Photography of Earth.

Figure 5.5: © Stuart Nock.

Figure 6.6: NOAA Photo Library/Wilson Bentley.

Figure 7.7B: NOAA Photo Library/Grant W Goodge.

Figure 7.8: adapted from Kiehl, JT and Trenberth, KE (1997): Earth's Annual Global Mean Energy Budget. Bulletin of the American Meteorological Society, #78, 197–208.

Figure 8.5A: © Daniel Peake.

Figures 8.5B, 13.5B: © John Lawson.

Figure 10.4: adapted from Bjerknes, J and Solberg, H (1922): Life Cycle of Cyclones and the Polar Front Theory of Atmospheric Circulation. Geofysiske Publikationer, #1, 3–18.

Figure 11.2: NASA Earth Observatory/NOAA.

Figure 11.8: based on Halpert, MS and Ropelewski, CF (1992): Surface Temperature Patterns Associated with the Southern Oscillation. Journal of Climate, #5, 577–593; Ropelewski, CF and Halpert, MS (1987): Global and Regional Scale Precipitation Patterns Associated with the El Niño–Southern Oscillation. Monthly Weather Review, #115, 1,606–1,626.

Figures 12.1A, 13.9: © Wagner Nogueira Neto/Ana Carolina Lacorte de Assis.

Figure 12.1C: © Robin Hogan.

Figure 12.3: © Andy Barrett.

Figure 12.6AB: © Robin Tanamachi.

Figure 12.6C: © Gerard Devine.

Figure 13.5A: © Claire Delsol.

Figure 15.4: NASA Earth Observatory/USGS Earth Observing-1.

Figure 15.5B: © Jane Shonk.

Figure 15.5C: © Peter Shonk; courtesy of Alec Murray.

Figure 16.1: adapted from Brohan, P; Kennedy, JJ; Harris, I; Tett, SFB and Jones, PD (2006): Uncertainty Estimates in Regional and Global Observed Temperature Changes: a New Data Set from 1850. Journal of Geophysical Research, #111, D12106, DOI: 10.1029/2005JD006548.

Figure 16.2: National Science Foundation, photography by Heidi Roop.

Figure 16.3: adapted from Petit, JR; Jouzel, J; Raynaud, D; Barkov, NI; Barnola, JM; Basile, I; Bender, M; Chappellaz, J; Davis, J; Delaygue, G; Delmotte, M; Kotlyakov, VM; Legrand, M; Lipenkov, V; Lorius, C; Pépin, L; Ritz, C; Saltzman, E and Stievenard, M (1999): Climate and Atmospheric History of the Past 420,000 Years from the Vostok Ice Core, Antarctica. Nature, #399, 429–436.

Figure 16.4: Pieter Tans, (NOAA/ESRL) and Ralph Keeling (Scripps Institute of Oceanography).

Figure 16.5: from IPCC Fourth Assessment Report, Figure 3.2.

Preface

Every time we step outdoors, or even look out of the window, we experience weather. Sometimes we are greeted by clear, blue skies; at other times we are faced with grey clouds. On some days, we feel the wind in our faces; on other days it can be completely calm. Some days are warm, some days are cold; some bring rain, some stay dry. Some even bring severe weather – heavy snowfall, freezing rain, tornadoes or dust storms. No two days of weather are the same, and the weather is always changing.

Weather is the complex interaction of heat and water within the atmosphere. Its power source, the Sun, provides massive amounts of energy in the form of sunlight, which heats the Earth's surface and sets the atmosphere in motion. We experience this motion as wind. Water in the atmosphere can exist in all three phases – solid ice, liquid water and water vapour – and as the air circulates, water in the atmosphere can switch between these three phases. Water vapour is invisible, but when it condenses to liquid or ice, it appears as cloud. These clouds can grow and bring rain, sleet, snow, hail, thunder and lightning.

Since the dawn of time, mankind has watched the weather changing from day to day and year to year. Despite this, being able to forecast the changing weather is a skill that has eluded us until only very recently. The vast numbers of calculations required to produce a reliable weather forecast have only been possible since the advent of the supercomputer. Modern technology has played a vital part in the development of the science of meteorology, not just in terms of forecasting, but also in improving our observations of the atmosphere, making meteorology perhaps one of the youngest of all the sciences.

Nowadays, meteorology is a very accessible science. TV weather forecasts are screened many times a day and a wealth of weather information is now freely available over the internet. Meteorologists also find themselves in the public eye a great deal more – every time they make a forecast, their science is put to public scrutiny. Their research is a constant, ongoing challenge to improve the ability to forecast the weather, not just through improving weather models, but also by improving understanding of the background science of meteorology, of how weather systems form and interact, and even improving techniques of observing the weather. Another area of increasing research is climate prediction: with rising global temperatures, attention is often turned to the climate scientist to understand how the future climate might vary.

From a very early age, we are exposed to weather. For some of us, this sparks an enthusiastic interest leading to a career in meteorology; others are encouraged to become dedicated amateur weather watchers. This book aims to provide a basic understanding of the science of meteorology for those keen to take it a little further. We begin by summarising the history of meteorological observations and forecasting (Chapter 2), then we look into how the weather is observed using a combination of surface observations and remote sensing techniques (Chapters 3 and 4). The next chapters introduce the basic physical science behind the weather (Chapters 5 to 9).

Four chapters then follow (10 to 13) that apply this science to the atmosphere to explain how various types of weather system evolve and dissipate. Then finally we move on to look at forecasting: how forecasts are generated from observations, and how the models we use to forecast the weather can be extended to generate predictions of future climate (Chapters 14 to 16).

Note: all terms highlighted in **bold** are defined in the Glossary at the end of this book.

The Earth from space. (Image: NASA Earth Observatory.)

1 Watching the Weather

Weather is one of those parts of our lives that we either love or hate. On a clear, sunny day we all rave about the weather being glorious, while on a dull, rainy day we rant about the weather being terrible. We often strike up conversations with friends and relatives and even strangers with an opinion about the weather – and as the weather is constantly changing, there is always something to talk about. From professional meteorologist to enthusiastic amateur, weather is one of those things of which we all have some degree of knowledge. The basic principle of making an observation and then a forecast is something we all do every time we look up at the sky and wonder if it is going to rain. It seems as if we belong to a world that is obsessed with the weather. But where does this obsession come from?

1.1 The Influence of Weather

Perhaps our obsession with the weather stems from the way that it can capture our imaginations. We may pay little attention to common, day-to-day weather events, such as clouds and rain, but when the weather produces something fascinating, such as a striking sunset, a stunning cloudscape or a picturesque snow scene, we often take note and reach for our cameras (see the figures in this chapter). The weather is often used as an evocative way of scene-setting in literature, and to set the mood of paintings. Indeed, the use of weather in the arts runs back thousands of years.

The weather influences our actions every single day without our even having to think about it. Like it or not, while we have the technology and knowledge to forecast the changing weather, there is nothing we can do to alter what is coming our way. Therefore, we must adapt our daily routine to the changes in weather from one day to the next. Before we leave the house in the morning on a rainy day, for example, we might choose to take a raincoat and an umbrella with us; on a cold day, we wrap up warm; on a hot, sunny day we might take our sunglasses. Once we have left our house, the weather can influence our journey to work: on a warm, sunny day, we may consider walking; on a rainy day, we may drive or take the bus. At lunchtime, we might consider a hot drink if the weather is cold; in hot weather, a cold, refreshing drink may be a more appropriate option. And when we arrive home after a day's work, the weather may influence whether we spend the evening outdoors, or spend it indoors in front of the TV.

It is not just individuals that can be affected by weather. The weather also has strong influences on businesses and industry. Agriculture, for example, is very dependent on the weather. Rain is required for the irrigation of crops, so a widespread lack of rain can result in crop failure and a significant drop in income to the farming community, with knock-on effects on food prices. Energy companies need to know about the forthcoming weather: if colder weather is on the way, there will be an increase

Figure 1.1 A sun pillar, seen here just after sunset. The vertical column of light above the Sun is reflected off horizontally aligned ice crystals in the clouds. (Photo: Rachel O'Byrne.)

in energy demand for heating. Also, nowadays, with so much energy being generated from renewable sources such as solar panels and wind turbines, the influence of the weather on the energy companies is perhaps greater than ever. The weather can even influence the stock put on display by supermarkets. A period of hot weather, for example, may inspire them to stock up on burgers and sausages for barbecues; a spell of wet weather may encourage them to display more umbrellas.

The weather can impact sporting events. Play usually stops during cricket and tennis matches when it starts to rain. Golf tournaments tend to be more resilient to rain, but the courses are quickly cleared if there is thunder and lightning about. The tactics of a motor racing team can be swung by the weather: forecasts of rain may affect the choice of tyres put on the cars. Often top racing teams will employ meteorologists to keep track of

Figure 1.2 The arc of a rainbow, seen in rain falling from a cloud over southern Malaysia. (Photo: Jon Shonk.)

the weather during a racing weekend. Horse racing is also affected by weather: the amount of recent rainfall affects the state of the ground on which the horses run.

Weather can affect our ability to travel, particularly when it turns adverse. For example, ice and snow on the roads can make driving conditions dangerous. The winter of 2009 to 2010, sometimes referred to as the Big Freeze, brought widespread snow and freezing temperatures to much of northern and western Europe from mid-December 2009 to late January 2010. This caused major disruption both to road and rail travel. It also played havoc with the aviation industry – one that is perhaps more sensitive to weather conditions than most. Snow and ice at airports can result in closures and mass cancellations of flights. The impacts of the freezing weather in early January 2010 had detrimental effects on the UK flight network, with most of the UK's major airports closed at some point, including both Heathrow and Gatwick. Western Europe also saw the effects of volcanic ash on aviation in April 2010. The eruption of the Icelandic volcano Eyjafjallajökull coincided with northwesterly winds across the Atlantic that spread its ash cloud over most of Western Europe, leading to cancelled flights and grounded aircraft.

In extreme cases, of course, weather can be devastating. It may be easy to forget on a fine, sunny day that, if conditions are right, intense weather systems can develop and bring strong, damaging winds and torrential rain. The impact of severe weather on humanity can be high, particularly if it strikes inhabited regions. Hurricanes, typhoons and tornadoes can cause widespread damage, and not just caused by the wind. Much of the devastation inflicted on the southern coast of the USA by Hurricane Katrina in August 2005 was due to flooding caused by storm surges. Trends in weather over longer periods can also be very damaging. For example, a long-term lack of rainfall in a region can lead to drought, which can have devastating effects on agriculture and cause widespread famine.

1.2 Weather Watchers

Whether fine or adverse, it always pays to know what weather is just around the corner. Fortunately, these days we have technology in place that can provide us with ample warning of severe weather. Weather forecasts are beamed into our lives on a daily basis

Figure 1.3 Liquid water at temperatures below 0°C can freeze on to cold surfaces. Spray from a nearby waterfall in Iceland has encased this grass in a crust of ice. (Photo: Michelle Cain.)

– every day, we are presented with weather maps that show both the current weather situation and how the weather is likely to change over the next few days. The weather is often a hot topic on TV, with any noteworthy weather events around the world usually grabbing a few minutes on a news programme, and occasional TV series commissioned describing the basics of meteorology. In combination, this media coverage is a strong influence on our knowledge and understanding of the weather.

In fact, we can gain a great deal of basic understanding of the weather by watching weather forecasts. Some TV forecasts still use maps of atmospheric pressure and weather fronts, and animate how the pressure field is predicted to evolve over the next few days. As a consequence, most of us are aware that areas of high pressure are usually associated with fine, settled weather, while low-pressure systems bring unsettled, rainy and stormy weather. We are also largely aware that the more closely packed the lines of constant pressure (known as **isobars**) are over an area, the stronger the wind will be; we probably also know that wind from the direction of the poles tends to bring much colder conditions than wind from the direction of the tropics.

These days, it is possible for any one of us to take an interest in the weather. The internet contains a vast wealth of meteorological information, both in terms of weather data, weather maps and openly available forecasts, enabling the enthusiastic weather watcher to constantly keep up to date with the current weather situation and keep track of how the weather might change over the next few days. It is also very straightforward to set up a weather observation station in our back garden, with thermometers and barometers readily available in the shops and online. Entire integrated weather stations, often combining temperature and pressure sensors with anemometers to measure wind, hygrometers to measure humidity and automatic rain gauges, are also now becoming cheaper to purchase. These are often sold with data loggers that can be connected to a computer, allowing us to create our own archives of weather records.

Even without such technology, it is still possible to remain very much connected to the weather. Measuring devices allow us to quantitatively observe many aspects of the weather. However, keeping track of what the weather is doing does not necessarily require such measurement. Indications of weather are often plainly visible in the sky: the distribution and shape of clouds can indicate what the current weather is like and also give warning of imminent change. Movements of the clouds can show the general direction of the wind. And, of course, our bodies have the ability to sense changes in temperature. Over time, if we observe the weather for long enough, it is possible to see patterns emerging, allowing simple forecasts of changes in weather to be made. In short, a lifetime of exposure to the changing weather has made us into a world obsessed with observing and forecasting the weather and, with meteorology becoming ever more accessible to the public, everyone can become a weather watcher.

2 From Seaweed to Supercomputers

It may be surprising that, despite the fact that the weather has been influencing us since the dawn of time, the science of meteorology remains fairly young, with most of the significant developments only happening relatively recently. To perform the two important steps required in meteorology, observation and forecasting, we need suitable measurement methods and computational ability – in other words, technology. Up until the last 50 years or so, meteorology has progressed fairly slowly. However, recent technological advances have been pivotal in enabling the science of meteorology to grow rapidly into the rich field of science that it is today.

2.1 The Age of Seaweed

The weather must have played a huge role in the lives of prehistoric man. In a time before thermally insulated houses and central heating, we would have had to adapt to and endure the swing of temperature from day to day and year to year. The weather would have affected the behaviour of animals, varying the supply of food. It would also have had an impact (as indeed it still does today) on crop yields.

Descriptive records of weather conditions exist from thousands of years BC. However, the first people to attempt to explain the science behind the weather were the Ancient Greeks. In a time when the Greek mathematicians such as Pythagoras and Euclid were laying down the basics of mathematics and physics, a few were trying to explain the complicated interactions that led to weather. Perhaps the first 'book' about weather was written by Aristotle in about 340 BC. Entitled *Meteorologica*, it gave explanations for many different types of weather, and was the accepted text on the matter for well over one thousand years. However, it had a number of inaccuracies – mostly brought about by Aristotle's attempt to explain everything using the interaction of the four classical elements: earth, air, fire and water. Nevertheless, some descriptions within it, such as the hydrological cycle, were actually quite accurate.

Weather watching and forecasting during this period remained qualitative and imprecise, mainly making use of natural indicators. Seaweed, for example, becomes dry and crisp when the humidity of the air is low, but becomes limp and moist when the air is humid and rain is more likely. Also, pine cones close when the air is humid and open when it is drier. Longer-range forecasts were based on, for example, the flowering of trees or the appearance of insects.

The use of such natural indicators led to the development of so-called **weather lore** – sayings about the weather that have been passed down through the generations. Even today, with the availability of forecasts on TV and online, a red sunset often inspires us to quote: 'Red sky at night, shepherd's delight; red sky in the morning, shepherd's warning.' This particular saying works only when weather systems approach from the

Figure 2.1 Red sky at night over Aberdeenshire, Scotland. A red sky at sunset indicates clearer skies to the west, hence finer weather on the way. (Photo: Jon Shonk.)

west – something that they normally do (Fig. 2.1). A red sky at sunset indicates clearer sky to the west, hence finer weather on the way; a red sky at sunrise suggests clearer sky to the east, and that the fine weather has passed. Many of these sayings, however, have been shown by current weather observations to be very unreliable. For example, 'oak before ash, we'll have a splash; ash before oak, we'll have a soak', linking summer rainfall to the order in which the oak and the ash trees come into leaf, shows no particular reliability. Indeed, in some languages, ash before oak is linked to a splash.

2.2 Early Meteorological Advances

The big step forward in the science of meteorology, made in the seventeenth century, was the invention of two instruments that remain at the heart of meteorological observations today: the **thermometer** and the **barometer**. These allowed weather observers to monitor temperature and pressure. The earliest thermometer, often attributed to Italian scientist Galileo Galilei, used a number of glass spheres of a fixed mass suspended in a fluid. Each sphere is designed to float at a slightly different temperature, meaning that the number of spheres floating at the top gives an indication of how warm it is. Even today, such thermometers are sold as **Galileo thermometers** (Fig. 2.2), although their slow response to temperature change limits their location to mantelpieces rather than weather stations.

The first barometer was invented by Evangelista Torricelli, another Italian scientist of Galileo's era, in 1644. His design was referred to as the **mercury barometer** – a design also still in use today (see left side of Fig. 2.3A). Torricelli filled a long tube of glass with mercury, and then inverted it into a bowl of mercury. He found that the height of the column of mercury settled at about 760 mm, and that it was atmospheric **pressure** pressing down on the surface of the mercury that was holding the mercury in place. From this, he was able to monitor atmospheric pressure by observing

Figure 2.2 A Galileo thermometer. Each glass sphere floats at a different temperature, indicated by the metal tag on the bottom. The temperature here is 21°C. (Photo: Jon Shonk.)

changes in the height of the mercury column.

Shortly afterwards, a number of Galileo's students set up a scientific society – the Accademia del Cimento – in Florence in 1657 with the aim of carrying out scientific investigations. One of the members of the Academy, Ferdinand II de Medici, deployed these early meteorological instruments at a number of locations across Europe and set up the first observational network. The scientists of the Academy also invented a number of primitive devices for measuring other atmospheric quantities, including a **hygrometer** that measured humidity by quantifying the amount of **condensation** onto a bucket of ice. Other scientific societies started forming in other European cities, including the Royal Society, founded in London in 1660, and the Académie des sciences, founded in Paris in 1666.

Further observational developments occurred over the next century. One of the shortcomings of early weather records was the lack of consistent scales, particularly for temperature. This was remedied independently by Gabriel Fahrenheit and Anders Celsius. Both the **Fahrenheit** and **Celsius** temperature scales are still used today, with the Celsius scale being more popular in Europe and the Fahrenheit scale preferred in the USA. Initially, the Celsius scale was set to run from 0°C to 100°C between the freezing and boiling points of water at standard atmospheric pressure; the Fahrenheit scale was set to run between 32°F and 96°F between the freezing point of water and human body temperature. Fahrenheit's other major addition to observational meteorology was the design of both the alcohol-in-glass and the mercury-in-glass thermometers in 1724 – again, a design that has remained largely unchanged.

The development of meteorology continued into the nineteenth century. The modern scheme for classifying clouds was proposed in 1802 by Luke Howard, using the Latin words cirrus, stratus, cumulus and nimbus to categorise all the different types of cloud. Shortly after, Francis **Beaufort** standardised measurements of wind using 13 categories spanning calm conditions (0) to hurricane-force conditions (12). This qualitative scale allowed wind speeds to be determined by observable features, such as waves at sea or movements of trees, in the absence of accurate measuring devices.

In the nineteenth century, the background science of meteorology started to become a subject of interest, with the physicists of the time beginning to understand the principles of **thermodynamics** and the flow of heat. These were investigated by a number of earlier scientists, but first summarised by Sadi Carnot. Out of this work, the laws of thermodynamics were developed by, among others, Émile Clapeyron, Hermann von Helmholtz, Rudolf Clausius and Lord Kelvin. In 1848, Lord Kelvin postulated the existence of a minimum possible temperature, known now as **absolute zero**. The **Kelvin** unit of temperature was later defined as the benchmark temperature unit, with the modern definitions of the Celsius and the Fahrenheit scales fixed to reference points on the Kelvin scale (slightly different from their original definitions).

The reporting of weather observations was hugely facilitated by the invention of the electric telegraph. Many inventors created telegraph systems through the early nineteenth century, with the 1836 system of Samuel Morse and his collaborators being the most successful. This allowed weather reporting to be much more 'real-time' – weather occurring in a given location could be reported instantly to a central location. This speed of communication was vital to the development of operational forecasting. Joseph Henry of the Smithsonian Institution set up a full observation system using Morse's telegraph in the USA in 1849. Using this network, daily reports of the weather nationwide were published in the *Washington Evening Post*.

2.3 The First Forecasts

Using the telegraph system, it was therefore possible to amass observations from stations all over Europe and plot weather maps. Early on, it was noted that stormy weather observed at one site was usually not a one-off event, but related to an area of low pressure that tracked across Europe. Forecasts of the passage of storms across Europe were provided during the 1850s by the Paris Observatory in France.

Robert Fitzroy made significant contributions to the world of weather forecasting. As well as being an officer in the Royal Navy, perhaps most famously captaining the HMS *Beagle* on the expedition that took naturalist Charles Darwin around the world, Fitzroy was a keen meteorologist. He collected weather data from a number of sites around the UK via telegraph and plotted weather charts. By comparing the current conditions with past weather maps, he was able to provide the first regular forecasts to be published in a newspaper – these appeared daily from 1860 in *The Times*. Fitzroy was also instrumental in organising weather data collection at sea by providing captains with meteorological instruments. His weather department was a forerunner of the current UK Met Office. Over the next 40 years or so, other national meteorological

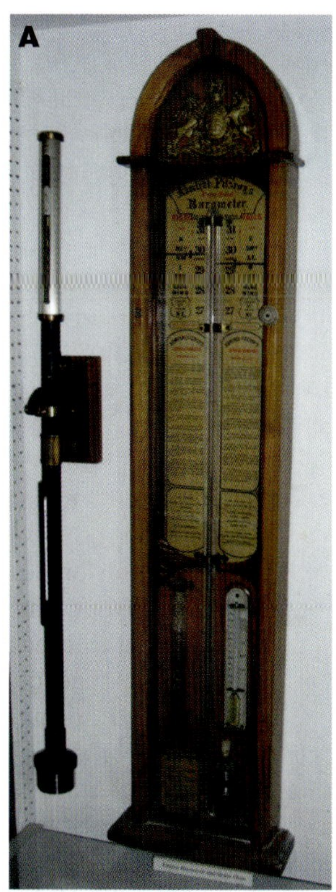

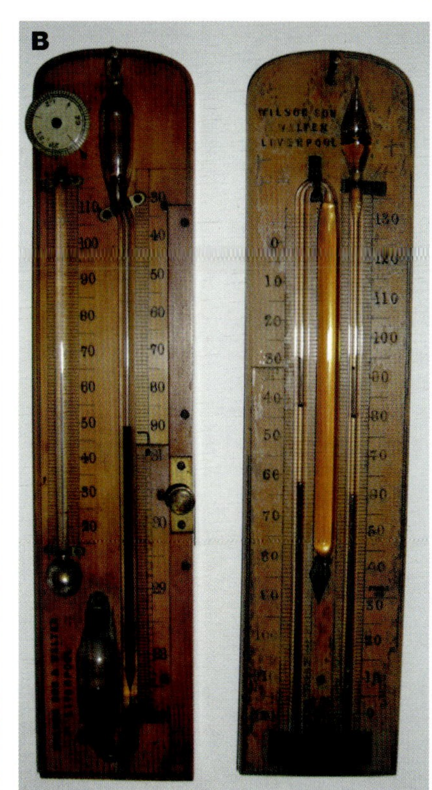

Figure 2.3 Some historical weather instruments. (**A**) Left: a mercury barometer. Right: a Fitzroy barometer, containing another mercury barometer, a thermometer and a storm glass. The wooden casing explains how to interpret the readings to give simple weather forecasts. (**B**) A pair of thermometers that record maximum and minimum temperatures. (**C**) An anemometer. The speed of rotation of the cups is converted to a wind speed on the analogue dial. (**D**) A tipping-bucket rain gauge. Rain falls into the funnel at the top and accumulates in one side of the copper bucket. When full, the bucket tips and the mechanism records a unit of rainfall. (Photos: Department of Meteorology, University of Reading.)

services began to appear, such as the Indian Meteorological Department, the US Weather Bureau (later to become the National Weather Service) and the Australian Bureau of Meteorology. An early analysis chart from the US Weather Bureau is shown in Figure 2.4.

Norwegian scientist Vilhelm Bjerknes made the next step towards modern forecasting in 1904. He made fundamental mathematical connections between thermodynamics and fluid mechanics, generating a set of mathematical equations that could, in theory, be used to numerically predict the weather. The first numerical forecast was produced several years later by Lewis Fry Richardson in 1922. Using a grid of weather data from 07:00 on 20 May 1910, he set out to create a six-hour forecast. Of course, he had to perform all the calculations by hand, as computing resources were unavailable at the time. The forecast turned out to predict unrealistically large changes in surface pressure because the equations he used allowed fluctuations to develop on a scale not observed in the atmosphere (modern forecasts mathematically filter out these unrealistic fluctuations). Even so, he had a vision of the first global numerical weather prediction, with a large number of forecasters

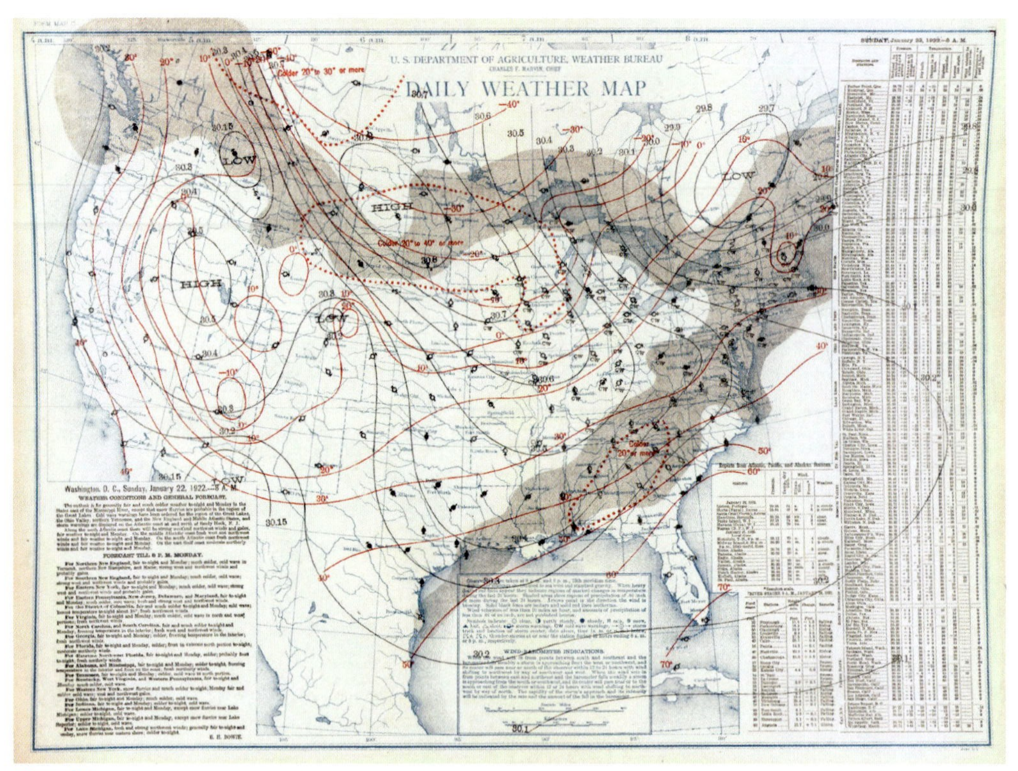

Figure 2.4 A map of weather over the USA on 22 January 1922. Contours of pressure (in inches of mercury) and temperature (in °F) are shown. (Image: NOAA Photo Library/US Weather Bureau.)

within a massive spherical room representing the Earth, with each forecaster calculating the weather at a given point based on the results found by the surrounding forecasters.

In the 1910s and 1920s, the earlier work on the passage of storms was advanced by a group of scientists based in Norway that included Vilhelm Bjerknes and his son Jakob. They developed a simple model describing the life cycle and evolution of **mid-latitude depressions** that used the now familiar concept of **fronts**, where cold air and warm air meet. Despite their limited observational techniques at the time, their description of how depressions develop was very accurate. Research has since added further understanding to the process, yet most elements of their so-called **Norwegian model** are still used today to explain how depressions form and develop.

2.4 Looking Up into the Atmosphere

Of course, forecasters needed information about the weather conditions not only at the surface, but also high up in the atmosphere. Early attempts to measure conditions at height used kites with instruments attached. However, as they had to be secured to the ground by ropes, they could never get particularly high into the atmosphere, as long ropes are very heavy. Manned ascents in hot-air balloons allowed some instrumented measurements – English meteorologist James Glaisher made many trips up through the atmosphere in hot-air balloons during the 1860s. However, this was often risky, with low levels of oxygen often causing observers to pass out during the flight. The development of the **radiosonde** during the 1920s made the routine collection of data aloft far less hazardous. The first true radiosonde was invented by French meteorologist Robert Bureau, although modern radiosondes follow the design of Russian meteorologist Pavel Molchanov.

The increase in aviation throughout this period necessitated the invention of the radiosonde, particularly during the Second World War, where in-depth forecasts and observations were required for military aircraft. The Second World War also led to the development of rainfall radar, another system that we still use today. **Radar**, an acronym for radio detection and ranging, was originally used as a means to detect incoming aircraft at airfields and military installations. Pulses of **radio waves** are emitted and reflect back off any solid objects, such as approaching aircraft. The radar operators also noted, however, that radio waves also reflected off areas of rain. After the war, a group of meteorological researchers investigated this effect further and noted that, if the wavelength of the emitted radio waves was tuned to about 50 mm, information about the raindrop size and location could be obtained. This became a valuable tool in the real-time monitoring of rainfall. An early rainfall radar image is seen in Figure 2.5

The final big step in weather observation came in 1960s, when the first weather satellite were launched. Earlier investigations using cameras mounted on rockets showed the immense amounts of information available from a space-based observation platform. NASA's '**TIROS-1**' (Television Infrared Observation Satellite) is considered the first successful weather satellite, returning pictures of clouds from space. Launched in 1960, it provided a wealth of images images of **cloud cover** around the world, such as in Figure 2.6. Subsequent TIROS satellite missions through the 1960s continued the research into the

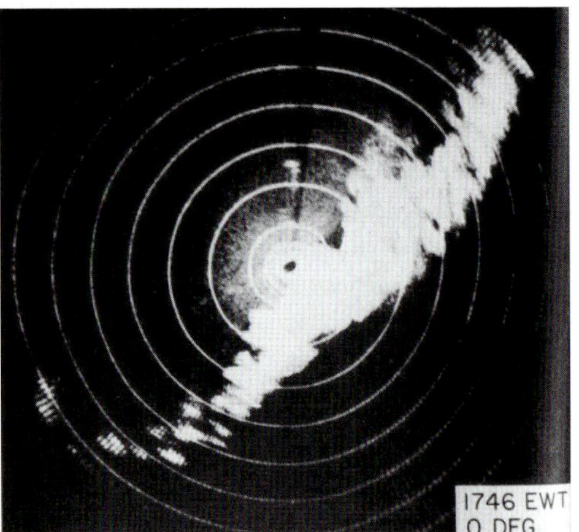

Figure 2.5 An early radar image, showing a line of storms over New Jersey, USA, taken on 16 July 1944. (Image: NOAA Photo Library.)

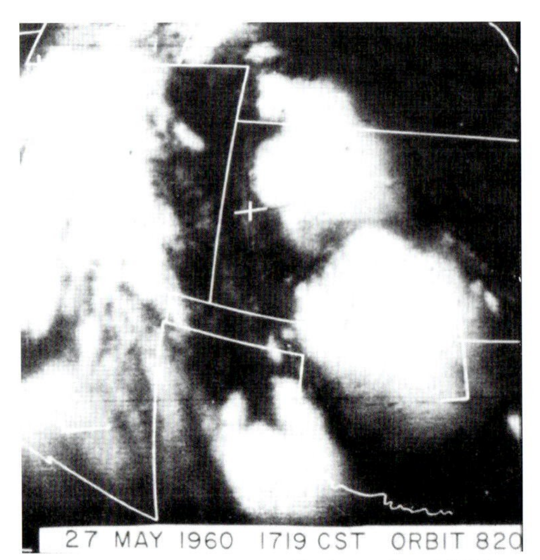

Figure 2.6 An early visible satellite image from TIROS, showing storm systems over the central USA on 27 May 1960. (Image: NOAA Photo Library.)

viability of weather observations from space. Nowadays, of course, the Earth is surrounded by a fleet of weather satellites constantly monitoring the atmosphere and feeding information back to Earthbound control centres, providing data that is used both for forecasting and research purposes.

2.5 Towards Modern Forecasting

Richardson's vision of a supercomputer consisting of a large number of human forecasters came several decades before the development of any computer capable of actually running a forecast model. Hence, despite these breakthroughs in the principles of numerical modelling of the weather, forecasting methods remained simplistic and based on pressure trends and comparison of weather conditions on a day with previous weather patterns. The demand for forecasts through the Second World War was high, with military aircraft requiring forecasts of wind patterns at height, along with cloud forecasts.

By 1947, the first computer capable of being programmed to run a weather forecast model was built by a group at the University of Pennsylvania in the USA. The ENIAC (Electronic Numerical Integrator And Computer) was the world's first fully electronic computer (Fig. 2.7). It used vacuum tubes and resistor technology and filled a whole room. A group of scientists, including American meteorologist Jule Charney and mathematician John von Neumann, performed a simple forecast of atmospheric pressure. It took about a day to produce the 24-hour forecast, and much of this time was spent manually dealing with the punch-cards required to feed the input into the computer. The results were very promising and paved the way for future forecast models. From this initial forecast, the world of meteorology has grown massively, with the development of faster and more powerful computers. Soon after the success with ENIAC, a number

Figure 2.7 ENIAC, the first fully electronic computer. It was used to run the first numerical weather forecast. (Image: US Army Photo.)

of national weather services began to run their own operational forecasts. As computers speeded up, more and more processes could be included in the weather models, giving increasingly detailed forecasts.

Developments in technology over the last 50 years or so have allowed massive advances in computing technology. Since about 1970, computing speed and power have risen exponentially with time. According to **Moore's Law**, the number of transistors that can be placed on an integrated circuit has been doubling every two years, and this growth in computing power continues. This means that much greater computational power can also be condensed down into a smaller space. For example, the computational power of a modern mobile phone that fits into the palm of our hand is much greater than the power of the early computers, which sometimes filled several rooms.

Computational power has also become cheaper. This has allowed advanced computing systems to become commonplace in research agencies and universities all over the world, allowing cutting-edge research into understanding the science behind the weather. Modern meteorological research encompasses three main areas: the development and improvement of instrumentation to increase our ability to observe the weather; analysis of the observed data to increase our understanding of the processes that take place in the atmosphere; and building better methods of modelling these aspects of the weather. Each of these areas of research spans a wide range of disciplines and, with improved computer technology, we can now analyse our data and model the weather in far greater detail than ever before.

One final chain of research branched off from the development of weather forecasts while computer power was still in the early stages of development. This was the realisation that these weather models could be modified and run over much longer periods. This opened the door to the principle of **climate modelling** and allowed scientists to investigate the possibility of future climate change.

3 The Weather Station

Before we can start to forecast the weather, we need a set of data describing the condition of the atmosphere at a given time. In forecasting circles, such data is referred to as **initial conditions**. We need observations of temperature, pressure, humidity, cloud cover, cloud type, wind speed and many other quantities, ideally at as many points on the Earth's surface as possible, and sampled throughout the depth of the atmosphere. Of course, it is impractical to observe the entire atmosphere at quite this level of detail. Even so, the modern global weather observation network consists of over 10,000 land-based weather sites, thousands of automatic weather stations mounted on buoys and ships at sea, hundreds of sites that monitor the conditions aloft, and an array of weather satellites in constant orbit around the Earth. Using data from all of these sources, we have adequate information to initialise our forecast. We begin our tour of the global observation network on the ground – at the meteorological observation station.

3.1 Surface Observations

Of all the types of weather observation, the first to develop was the land-based surface observation station. These have been around for centuries and were, in the past, manned with observers making records of the weather using sometimes very simple meteorological instruments. There were no set standards on the location and siting of observation stations, so weather readings were sometimes variable in quality. Even so, there were enough weather data archived in various historical records from around the Midlands of England to allow English scientist Gordon Manley to create a temperature record stretching all the way back to 1659. This series is known as the **Central England Temperature** series, and is the longest observational record of climate in existence. Since Manley completed the series in 1953, it has been updated every year. The modern UK operational surface observation network consists of several hundred stations. Many of these are fully automated and return their measurements electronically, but some still employ an observer to make measurements in the traditional way.

A great number of countries in the world now have networks of surface observation stations. Figure 3.1 indicates the typical global coverage of surface observations. But, if the data they report is to be useful, it is important that international standards are set so that data from one country is consistent with data from the next. The **World Meteorological Organization** (WMO), a branch of the United Nations, defines these standards. For a global set of weather observations to be useful to a forecaster, it is much better if they are all made at exactly the same time. Therefore, irrespective of time zone, the meteorological world sets its clocks to **Universal Time Coordinated** (UTC), chosen by convention to be synchronised with **Greenwich Mean Time**.

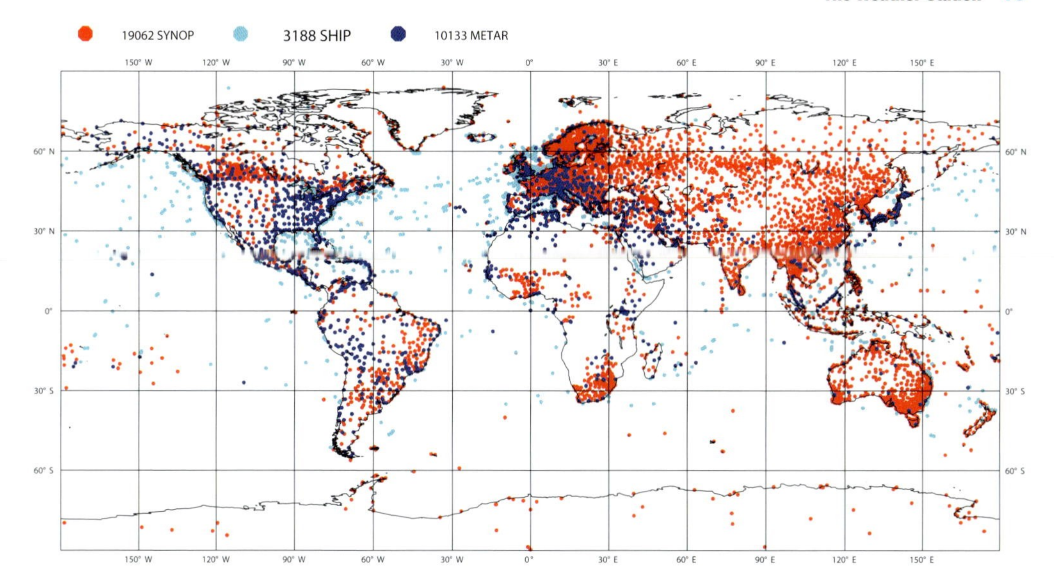

Figure 3.1 Each coloured spot shows the location of one of the 32,383 surface stations that provided a weather report at 00:00 UTC on 22 November 2011. Note the uneven distribution of the observations over the globe. (Data: ECMWF.)

That way, making observations worldwide at 00:00, 06:00, 12:00 and 18:00 UTC gives global snapshots of the weather every six hours.

Often, however, weather observations are made much more frequently than six-hourly. In operational stations, readings are normally made every hour throughout the day. These include readings of temperature, pressure, humidity, wind speed, wind direction and many more. Such observations are referred to as **synoptic observations**. Some stations also make **climatic observations**. These are measurements that are made once a day (usually at 09:00 local time) and summarise the weather over the course of the preceding 24-hour period. Climatic observations include maximum and minimum temperatures, total rainfall accumulation and hours of sunshine.

When made, these observations are relayed back to the national weather services, where the data is collated and shared with the world. Again, so that the observations can be interpreted by any meteorologist anywhere in the world, the observations are reported using a coding system managed by the WMO. For a weather agency in a particular country to be able to generate a forecast, it needs weather data from all over the world. The observations collated by the national services are then gathered and archived at one of the WMO's World Meteorological Centres.

3.2 Inside the Stevenson Screen

Even with the onset of modern technological advances, the instrumentation found on weather stations managed by an observer has, in many cases, undergone little change in the

Figure 3.2 (**A**) A modern Stevenson screen. (**B**) The standard arrangement of thermometers inside consists of dry-bulb and wet-bulb thermometers (mounted vertically to the left and right respectively) and maximum and minimum thermometers (mounted horizontally, top and bottom). (Photos: Department of Meteorology.)

last century. This is mainly for consistency: for ease of comparison of current readings with older readings, it is far more convenient if the same conventions are followed. Again, the WMO sets standards on measuring devices. It also has a set of standards for the location of weather stations. Ideally, they need to be as far away as possible from any obstacles that may affect the background meteorological conditions – certainly nowhere near buildings or trees that can block the wind or shelter the site from rain.

A signature feature of many a weather station is the **Stevenson screen** – a ventilated box to shelter instruments from sunlight while still allowing air to freely pass over them (Fig. 3.2A). Very early measurements of temperature were made with the thermometer exposed to full sunshine. However, it was later realised that exposing a thermometer in such a way results in readings that measure the temperature the Sun can heat the thermometer to rather than the air temperature, which is of far more interest. This inspired British engineer Thomas Stevenson in 1864 to design the Stevenson screen. Modern Stevenson screens are still based on Stevenson's original design and are made of louvred panels that are painted white. The standard height of a Stevenson screen, as determined by the WMO, is 1.5 m above the ground.

A Stevenson screen usually contains four thermometers: a **dry-bulb thermometer**, a **wet-bulb thermometer**, a **maximum**

thermometer and a **minimum thermometer** (Fig. 3.2B). The thermometers found in a Stevenson screen are usually mercury-in-glass, still following the design popularised centuries ago by Fahrenheit. They consist of a small reservoir of mercury connected to a fine glass thread. As the mercury in the reservoir warms, it expands along the thread. In many locations (particularly near the poles), alcohol-in-glass thermometers are preferred, as alcohol has a much lower freezing point (mercury freezes at −40°C). All temperatures measured on a weather station are usually quoted in degrees Celsius.

The current temperature is read off the dry-bulb thermometer every hour. The maximum and minimum thermometers measure the temperature extremes reached over a 24-hour period and are read once a day. Both maximum and minimum thermometers follow the same design as a normal thermometer, but with slight modifications. The maximum thermometer has a constriction in the thread that allows mercury to expand out of the reservoir as it warms up, but not to return as it cools. Minimum thermometers tend to be of an alcohol-in-glass design, and contain a pin within the alcohol column that is dragged along the tube towards the reservoir by the meniscus as the temperature falls. Records of maximum and minimum temperatures, while not particularly useful to a weather forecast, tend to be of great public interest: for example, the severity of a spell of either very hot or very cold weather is often determined in the media by extreme temperatures.

Many weather stations also place a second minimum thermometer on the open grass. As the ground heats and cools much more rapidly than the lower atmosphere, the daily swing of temperatures at grass level can be much greater. The overnight minimum temperature recorded here (referred to as **grass minimum**) is usually several degrees colder than the minimum temperature recorded in the Stevenson screen. A **ground frost** is said to occur if the grass temperature falls below 0°C overnight; an **air frost** occurs when the temperature in the screen also falls below 0°C.

Measurement of **humidity** is also made within the Stevenson screen. Traditional humidity measurements were made with **hygrometers**, which measured humidity in terms of the length of a bundle of hairs (hair becomes longer in humid conditions). However, these were tricky to calibrate. Modern humidity measurements usually use a **psychrometer**. This consists of a pair of thermometers, one of which has its bulb exposed to the air, the other of which has its bulb kept continually moist. As air passes over the wet bulb, it evaporates water from the bulb, causing it to cool. Comparison of the temperatures measured by the dry-bulb and wet-bulb thermometers can then be converted to humidity using a slide rule or a set of psychrometric tables. Humidity is expressed in terms of **relative humidity** – the amount of water vapour in the air as a fraction of the maximum water vapour that the air can hold at a given temperature, usually expressed as a percentage. Humidity is introduced in more detail in Chapter 6.

3.3 Outside the Stevenson Screen

Daily weather observations must also report the wind – both its speed and its direction (see Fig. 3.3). The measurement of wind direction still involves wind vanes, a design largely unchanged from the weathercocks found on church steeples, although modern vanes

Figure 3.3 Instruments for measuring wind: a traditional cup anemometer (left), a wind vane (centre) and a sonic anemometer (right). The sonic anemometer measures wind speed by emitting pulses of sound waves and timing their journey from transmitter to receiver. (Photo: Department of Meteorology.)

tend to be smaller and freer to move with the wind. Wind direction is reported as a compass bearing, usually to the nearest 5° or 10°, with bearings increasing clockwise from the north. Meteorological convention declares that wind directions are reported as the direction the wind is coming from rather than the direction it is going to. Hence a westerly wind, with a bearing of 270°, is moving from the west towards the east (in an eastward direction), while an easterly wind has a bearing of 90° and moves from the east towards the west (in a westward direction). Recent changes in wind direction are also sometimes reported, as changes in wind direction can indicate potential changes in the weather. A wind that is changing in a clockwise direction is said to be **veering**; one that changes in an anticlockwise direction is said to be **backing**.

The speed of the wind is measured using an **anemometer**, usually consisting of three cups attached to a vertical spindle that is free to rotate with the wind. The speed of rotation of the cups can then be electronically or mechanically converted to a wind speed. Wind speeds are quoted in either metres per second ($m\,s^{-1}$), or nautical miles per hour (**knots**; kt), although these are usually converted to miles per hour (mph) or kilometres per hour ($km\,h^{-1}$) for the general public. It is important to ensure that the flow of the wind to the anemometer is not affected by any tall obstructions. For this reason, anemometers are usually mounted at standard heights of 2 m or 10 m above the ground.

Air pressure is measured using a barometer. One of two types of barometer is used in weather observations. The first is the mercury barometer – again, a design largely unchanged since its invention by Evangelista Torricelli. Nowadays, however, **aneroid barometers** are more popular – and, indeed, more practical. These use a partially evacuated metal capsule that expands and contracts with changes in air pressure. The capsule is connected to a mechanism linking it to a dial display. Aneroid barometers also have the added advantage of size: the capsule can be made very small if necessary. Air pressure is reported in **hectopascals** (hPa) or **millibars** (mb), where 1 hPa equals 1 mb. Some home barometers measure pressure in inches or millimetres of mercury (inHg or mmHg) – the height of mercury that could be supported by the atmospheric pressure in a mercury barometer.

Air pressure is very dependent on altitude (see Chapter 5). The pressure recorded at a weather station is referred to as **station pressure**. However, if plotted on a weather map, the changes in pressure caused by the differing altitudes above sea level of weather stations would swamp the subtle differences across weather systems. For this reason, pressure is converted to what it would be if the weather station were to be lowered to sea level – its equivalent **mean-sea-level pressure**.

Changes in pressure over time, known as **pressure trends**, are an important indicator of changes to the weather; hence, it is usual to report how the pressure has changed over the last three hours. Many weather stations contain a **barograph** to determine pressure trends (Fig. 3.4). A barograph uses the same aneroid capsule as modern barometers, but with the dial replaced by an arm that moves up and down with changing pressure. A drum is set to slowly rotate over the course of several days, allowing a pen mounted on the end of the arm to trace out the changing air pressure.

Rainfall accumulation at a weather station is reported as the depth of water that falls on a given area, expressed in millimetres or inches. This accumulation is measured using a rain gauge, as seen in Figure 3.5, consisting of a cylindrical bucket with a funnel that directs the rain into a container. Each day, the rain gathered in the container is transferred to a narrow measuring cylinder that allows the volume of rain to be converted into a rainfall amount, quoted to the nearest 0.1 mm.

Figure 3.4 This barograph measures pressure using an aneroid capsule, which is linked mechanically to an arm with a pen on the end. The drum on the left rotates slowly, allowing the pen to trace out changes in pressure on the chart. (Photo: Department of Meteorology.)

Figure 3.5 (A) A standard rain gauge assembly. (B) The top of the gauge is a copper funnel (right), which directs water into a glass container inside (left). The rainfall collected in the container is measured using the calibrated cylinder (bottom). (Photos: Department of Meteorology.)

The timing of rainfall events throughout the day can be measured using a device of similar design to the barograph, but with the aneroid capsule replaced by a float in a container. As the rain falls into the gauge, the float then rises, pushing a pen up the chart attached to the rotating drum. An alternative to this is the **tipping-bucket rain gauge**. These replace the calibrated container with a pair of containers balanced on a pivot. Each container is designed to hold a fixed amount of rain (often the equivalent of 0.2 mm of rainfall). When it fills, it tips over and empties, allowing the container on the other side of the pivot to fill. The timing of these tips can be recorded to give an indication of the timing of rainfall.

As with most of the instruments, care must be taken when siting a rain gauge to ensure that there are no surrounding obstacles that might stop the rain falling into the gauge. Snow and hail are also counted in the daily rainfall record, although they must first be melted before they can be measured. A snow depth of 12 mm gives a rainfall reading of about 1 mm. Some rain gauges are also heated so that any frozen precipitation is melted straight away.

3.4 Watching the Skies

There are several more properties of the weather reported during a routine weather observation. Those listed so far can be read off dials and scales as a number. The rest, however, rely on perhaps the most frequently used weather observation tool: the human eye. These observations are much more dependent on the judgement of a given observer – and for this reason it is important that observers are properly trained so that the judgements of many observers for a given weather situation are as similar to one another as possible.

Human judgement plays a significant role in the daily reporting of sunshine duration. The standard device to measure this is the **Campbell–Stokes sunshine recorder**, designed in the late nineteenth century by John Campbell and modified to its current design by George Stokes (Fig. 3.6A). It consists

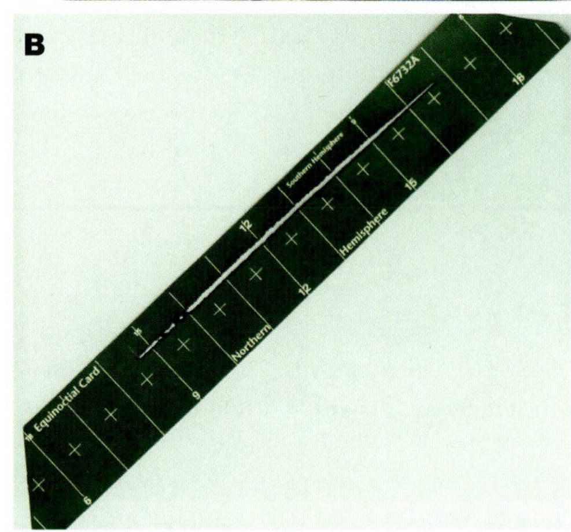

Figure 3.6 (**A**) A Campbell–Stokes sunshine recorder is a glass sphere that focuses the Sun's rays onto a card at the back. (**B**) The burn card from a mostly sunny day (03 October 2011). (Photos: Department of Meteorology.)

of a glass sphere that focuses the Sun's rays onto a strip of card mounted behind it (Fig. 3.6B). While the Sun is shining onto the sphere, it burns a track along the card as it moves across the sky. This card is replaced every day and gives a record of both the duration and the timing of sunshine throughout the day. The card is calibrated in hours and the length of the scorch marks can be added up to give total sunshine duration in hours. Human discretion is required, however, to determine the length of each individual burn and to add them together.

The human eye also remains the best instrument for determining visibility – the furthest distance at which objects can still clearly be seen. In clear, dry conditions, a visibility of several tens of kilometres is not unreasonable; in thick fog, visibility can be as small as a few tens of metres. Reference points of varying distance from the weather station are listed, from nearby buildings and trees to hills and features on the horizon. **Visibility** is determined by how many of these listed features can be seen. It is stated in kilometres.

For sites that report the weather hourly, it is also important to take an observation of the general state of the current weather, and to make note of what has happened to the weather in the past hour. The WMO's guidelines contain 100 different options to describe the current weather and 10 options for the past weather, each one with a different two-digit or one-digit code. There are some very subtle differences between some weather types, so only a human observer could realistically make this type of observation.

Finally, the observer must take observations of both cloud cover and the cloud types present at the time of observation. Cloud cover is simply determined by looking up to the sky and determining the fraction of it that is covered by clouds. This fraction is determined in eighths, more commonly referred to as **oktas**. A cloud cover of 0 oktas indicates clear sky; 4 oktas indicates that the sky is half-covered with cloud; 8 oktas indicates overcast conditions. Cloud cover is rounded to the nearest whole okta.

Reports of cloud type still use the nomenclature first laid out by Luke Howard in 1802. He divided clouds into four broad categories: cumulus, cirrus, nimbus and stratus. **Cumulus** clouds had flat bases and billowed upwards; **cirrus** clouds were light, wispy and fibrous; **stratus** clouds were flat and fairly uniform; **nimbus** clouds brought rain. He also identified clouds that satisfied more than one of his definitions, and labelled these using pairs of his cloud types, such as 'cirro-stratus' and 'strato-cumulus'.

These days, clouds are divided into three types – high-level, mid-level and low-level – and are still described using Howard's classifications. Cumulus is still used for fluffy, flat-based, fair-weather clouds composed of liquid water. Extra descriptors are sometimes used to further define subsets of a cloud type (see Fig. 3.7). For example, short cumulus clouds are referred to as **cumulus humilis**, while tall, deep, cumulus clouds are referred to as **cumulus congestus**. When a cumulus cloud takes on a wispy cap of ice, it becomes known as a **cumulonimbus**, which can bring heavy showers with thunder and lightning. While these two cloud types can extend through all three of the level categories, they are referred to as low-level.

Other low-level clouds are stratus, nimbostratus and stratocumulus. Stratus cloud

A
B
C
D
E
F

Figure 3.7 opposite Cloud gallery: (**A**) cirrus, often referred to as mare's tails; (**B**) a layer of stratocumulus viewed from above; (**C**) a layer of altostratus partially obscuring the Sun, with scattered altocumulus below; (**D**) contrail cirrus, a type of cloud that forms in the wake of jet aircraft; (**E**) stratus enveloping the top of a hill; (**F**) towering cumulus. (Photos: Jon Shonk (A–C); Mike Blackburn (D, F); Peter Smith (E).)

appears as a uniform sheet of cloud, usually consisting of liquid water droplets and found in the lower atmosphere below a height of 2 km. Thicker, precipitating layers of stratus are referred to as **nimbostratus**. Bumpy layers of cloud containing individual cells are referred to as **stratocumulus**.

Cirrus clouds are thin, wispy, ice clouds found very high up in the atmosphere. They can appear as parallel streaks of cloud or masses of tangled streaks. If the cirrus clouds thicken to form a widespread sheet, they become known as **cirrostratus**. Ripples in a sheet of cirrostratus can generate cells of **cirrocumulus**, often referred to as 'mackerel sky' and the rarest of the ten main cloud types. If similar clouds form at lower levels, they become known as altostratus and altocumulus respectively. **Altocumulus** consists of liquid droplets; **altostratus** can contain a combination of ice and liquid water. The differences between these types of cloud can be very subtle and usually only determinable by opacity: the high-level cirro-form clouds tend to be much more transparent than the mid-level alto-form clouds. Cirrostratus and cirrocumulus can also generate interesting optical effects, while altostratus and altocumulus tend to be thick enough to block the Sun and cast shadows.

3.5 Automatic Weather Stations

With the thousands of staffed surface weather observation stations across the world, it may seem that we should have easily enough information to enable us to build our snapshot of global weather. However, as we saw in Figure 3.1, the distribution of stations is far from uniform. Staffed weather stations tend to be located in places that are inhabited. In other words, there are vast swathes of the Earth's surface with very sparse coverage: for example, manual stations are impractical in the centre of the Sahara Desert or the frozen wastes of Antarctica. Poorer countries tend to have sparser networks of weather observations. The sea is also poorly covered by surface observations: traditionally, observations could only be provided by suitably equipped ships.

Modern technology has, however, now led to an increasing number of weather stations becoming automatic. An automatic observation station costs drastically less to run than a manual one, as the only human intervention required is occasional maintenance. They can be deployed in remote areas, filling gaping holes in the observation network, and can also be mounted on buoys and set to drift in the ocean (Fig. 3.8). Currently, there is a network of over 1,000 buoys measuring both

Figure 3.8 An automated weather buoy. This one is 10 m across – note the size of the scientists on board. (Photo: NOAA.)

the weather above the ocean and the properties of the ocean beneath.

Many of the quantities observed manually at a weather station can be fully automated using electronic sensors. For example, electronic temperature sensors consist of a coil of wire called a **thermocouple**. Change in temperature alters the electrical resistance of the wire, so if a small constant current is passed through it, the voltage across it (which varies as temperature changes) can be measured and logged. Humidity sensors can be built on the same principle if a material is used whose resistance changes with humidity. Alternatively, a pair of electronic sensors can be set up as a psychrometer.

It may therefore seem that automated weather stations are the answer for the future of weather observation, as they are cheaper to run and can be deployed anywhere as required. However, there remain a number of meteorological quantities that are still a significant challenge for the automation process – mainly the observations that require some degree of human judgement. Designing an electronic device that can take pictures of the sky and then analyse the cloud types would be a massive challenge, as would creating a network of sensors to identify which of the WMO's 100 descriptions of current weather is most appropriate. Maintaining these types of observations at least at some weather stations is important, as they are useful in determining the overall weather conditions.

Another inhibitor to automation, as mentioned previously, is consistency. Changing an entire network of observation stations to automatic sensors could cause sudden steps in recorded climatologies such as the Central England Temperature record. For example, automation of sunshine recording is straightforward – a device called a **solarimeter** is often used in weather stations to measure the amount of **radiation** from the Sun reaching the Earth's surface. It would be a trivial exercise to evaluate data from a solarimeter to determine when the Sun was shining. However, making records of sunshine from a solarimeter agree with those from a Campbell–Stokes sunshine recorder would be a much greater challenge.

4 Gauging the Atmosphere

Surface observations are, of course, only part of the observational story. Our hourly observations may provide us with a wealth of information at our weather station, but there is a great deal more weather out there to be sampled, both elsewhere between the weather stations and, of course, throughout the depth of the atmosphere. To give us enough information for the initial conditions of our forecast, we need information about the weather at height and also, as much as possible, information on what is going on between the weather stations – particularly over the oceans, where surface observations are sparse.

4.1 Measuring the Upper Air

A great deal of information about the coming weather that could affect us down on the Earth's surface is contained aloft, high up in the **troposphere** (the lowest layer of the atmosphere where most weather occurs). So, simply generating our weather forecast from surface observations could easily result in changes in the weather situation being missed. However, observing the so-called '**upper air**' is an expensive business, and one that is usually constrained to national weather centres and research organisations.

Ever since its invention by Bureau in 1928, the radiosonde has been the preferred method of measuring the upper air. Basically, the radiosonde is an automatic weather station, but miniaturised into a box no bigger than a small cereal packet. The typical mass of a radiosonde is about 200 g, although much lighter ones are available. Packed into this small space are three instruments: a thermometer, a hygrometer and a barometer. The thermometer is a tiny **thermistor** or wire coil whose resistance varies with temperature. Humidity can be sensed using a capacitor, made up of two plates with a dielectric material between them. As humidity varies, the properties of this material change, allowing humidity to be measured in terms of the capacitance. The barometer is usually a tiny aneroid capsule. These sensors are used on account of their inexpensive cost and their robustness at a wide range of temperatures: in flight, a radiosonde could be subjected to temperatures between 20°C and –70°C. For this reason, it also needs a very resilient, high-capacity battery to power the sensors. Finally, the packet contains a small transmitter that encodes the readings from each of the sensors and transmits them back to the ground in real time via radio. More advanced radiosondes also contain Global Positioning System (GPS) sensors that allow the location of the radiosonde to be tracked, giving information about wind speed and direction.

The development of the radiosonde has been an exercise in miniaturisation, made possible by the same technological advances that have helped in the advancement of automatic weather stations on the ground, as seen in Figure 4.1. Original designs of radiosonde were much heavier and bulkier and often relied on mechanical systems to switch between the sensors during the ascent.

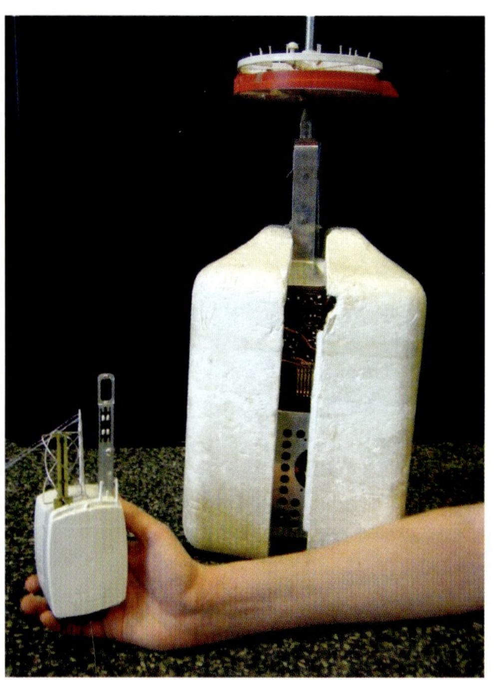

Figure 4.1 Modern radiosondes now fit in the palm of a meteorologist's hand – much more compact than earlier designs. The large radiosonde at the back dates from the 1970s. (Photo: Department of Meteorology.)

Figure 4.2 A radiosonde launch. The assembly consists of a radiosonde (the small white box at the bottom), a parachute (the red piece of plastic) and a helium balloon, which usually measures about 2 m across at launch. (Photo: Department of Meteorology.)

Modern radiosondes have become fully electronic, eliminating any problems caused by mechanical systems freezing high up in the atmosphere.

The routine observation of the upper air is a far more recent development than surface observation. The radiosonde started to make its mark on the meteorological world during the Second World War in the 1940s, when upper air data was required to forecast conditions aloft for fighter pilots. By the mid-1950s, radiosondes were widely in operational use. Nowadays, they are launched twice a day at 00:00 and 12:00 UTC from a network of upper air stations across the world (some places launch four a day). The radiosonde itself is connected by a cord to a latex balloon filled with helium or hydrogen (Fig. 4.2). To lift its payload, the balloon needs to be large enough – weather balloons are typically 2 m in diameter when fully inflated before launch. A balloon of this size can carry the launched radiosonde up to heights of 25 km to 30 km. At this height, as the air pressure is much lower (see Chapter 5), the balloon may have increased in size to over 10 m across. Eventually, the pressure difference becomes too great and the balloon bursts. The cord connecting the radiosonde to the balloon also contains a parachute – this is deployed, and the whole package drifts back down to the ground.

Data returned from a radiosonde ascent contains information about the state of the atmosphere throughout its depth and is referred to as a **profile** or a **sounding**. In essence, this can be considered to be a vertical sample of the atmosphere at the time of launch. In reality, of course, the measured profile will not be for the atmosphere directly above the launch site: the radiosonde is free to drift with the wind. Over the course of a 25 km ascent, the weather balloon can often travel as far as 100 km horizontally. The GPS sensors on the radiosonde allow the observers on the ground to keep track of the location of the radiosonde. Even so, they usually travel far too far for it to be worthwhile to retrieve them; indeed, many simply fall into the sea never to be found again. This single-use approach to upper air monitoring is one of the reasons why it is such an expensive process. Radiosondes can be bulk-produced, but every one needs to be individually calibrated to the highest standards for its readings to be useful. Even in bulk, the cost of a radiosonde, a weather balloon and the helium to fill it is likely to be in excess of £100. Despite their cost, however, launching radiosondes remains much cheaper than operating an aircraft or a satellite, and provides much more detailed information, returning measurements every 5 m or so.

Around the world, there are currently fewer than 1,000 sites that routinely launch radiosondes and report back on the upper air. As the weather state in the upper air contains far less local variation, a sparser observation network is not necessarily a problem. Even so, the upper air sounding network suffers similar problems to the surface network: global coverage is not uniform, with remote land areas having little coverage and few radiosondes being routinely launched at sea. This is far from ideal, as a great many weather systems form over the oceans, with fluctuations in upper-level winds giving warning of their development. Commercial airlines can assist in the problem by collecting data from the upper air. As they fly, they regularly monitor the weather conditions at cruise heights (10 km to 12 km). However, this is only a partial solution to the problem; the tracks of aircraft do not sample the whole globe – the areas they can provide data for tend to be limited to specific flight tracks. They also provide data only at a single level as they fly, only giving a full profile when they take off and land.

4.2 Radar and Lidar

So we have a set of measurements of the state of the atmosphere at a number of individual points. Whether drawing weather maps or defining the initial conditions for our forecast, we are still making an important assumption: that the properties of the atmosphere vary smoothly between these points. In the case of pressure fields, for example, this is a reasonable assumption – we do not observe fluctuations in pressure on a scale similar to the separation of weather stations. However, some quantities vary on a much more local scale, with fluctuations in these quantities being missed if they pass between weather stations. The most important of these are the presence of clouds and rain – a very heavy shower has the potential to cause local flooding, yet it is perfectly possible for its track to never pass over an operational rain gauge.

Fortunately, we also have **remote sensing** in our arsenal of modern observation techniques. Remote sensing is the monitoring of weather over a vast area from a single point, enabling gaps in the surface observation network to be filled in. The use of radar to measure rainfall is a widespread and well-known technique, with radar images of recent rainfall often making an appearance in TV weather forecasts (Fig. 4.3). The evolution of weather radar out of the types of radar used in the Second World War to monitor incoming enemy aircraft took a few decades but, by the early 1980s, weather radar installations were becoming integrated into networks, giving indications of present rain distributions over entire nations. Most developed nations now have dense networks of weather radars.

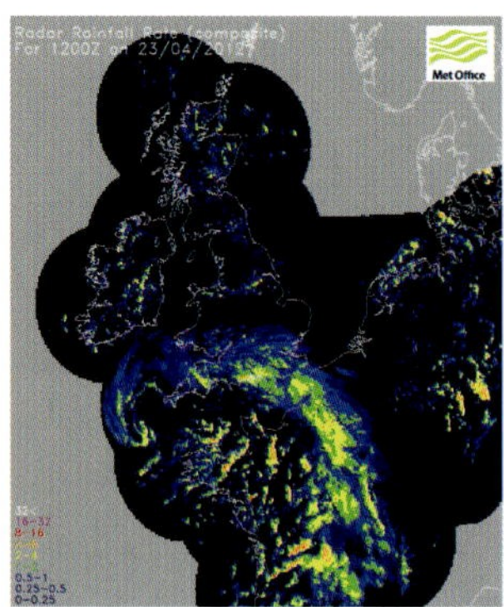

Figure 4.3 A radar image showing rainfall from a low-pressure system extending over southern England and northern France on 23 April 2012. (Image: Met Office.)

A modern radar site consists of an antenna in the form of a parabolic dish (Fig. 4.4). Usually, the dish is housed inside a dome to protect it from the weather. The radar is sited at the top of a hill to maximise its view of the surrounding area. The process of a radar scan is, in principle, exactly the same whether used to observe precipitation or incoming aircraft – a pulse of radio waves is transmitted at a fixed wavelength (Fig. 4.5). When this reaches a target, it reflects back off it and the receiver site detects the returned pulse. As radio waves travel at the speed of light, the delay between the transmitted and received pulse can then be converted to a distance. To detect rain, the radio waves are tuned to wavelengths of 50 to 100 mm (microwaves).

Figure 4.4 Operational radars are housed within a dome. This radar is at Dean Hill, and is part of the UK operational radar network. (Photo: Met Office.)

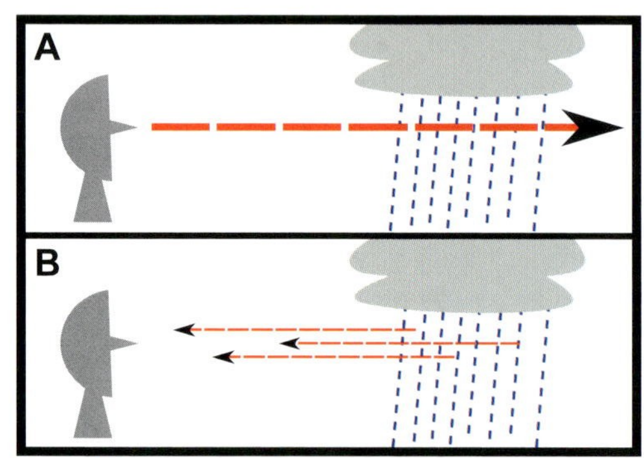

Figure 4.5 The basic principle of a radar retrieval. A pulse of microwave radiation is transmitted from the radar (**A**). A small fraction of this is reflected from raindrops, and this signal is received at the radar (**B**). The distance and intensity of the rainfall can be inferred from the strength and timing of the return signal.

The intensity of the rainfall can be determined by comparing the strength of the received signal and the transmitted signal. The ratio of the strength of these, when corrected for range, is referred to as the **radar**

reflectivity. In an area of more intense rain, there will be a greater number of large drops, hence reflecting a stronger signal. Radar reflectivity is then converted to rainfall rate. Radar is often calibrated using rain gauge measurements – although it is important to calibrate it over a long period of time, as local variability in rainfall accumulations over a point can be great, while the radar gives averages over areas of 1 km^2 or more.

A raw radar image is likely to require some correction before it is in a final form suitable for inclusion in our initial conditions. Rainfall is not the only thing that radar will detect. It also detects other precipitation forms, such as snow and hail. However, it can also detect targets that are not caused by precipitation, and hence it is important to remove any return signals that are caused by such targets. So-called '**ground clutter**' must be removed – this consists of returns from hills, buildings, trees or other solid, immovable objects. In some instances, radar images can also contain strong retrievals caused by **anomalous propagation**. This is a similar effect to a mirage seen on a hot, sunny day. A low-level **temperature inversion** can bend the beam back towards the Earth's surface, giving a strong return off the ground.

The curvature of the Earth also presents a problem for radar retrievals. As they travel through the atmosphere, radio waves are refracted towards the Earth's surface, but by the time a pulse has travelled over the maximum detectable range of a radar (250 km or so), it is now sampling the weather several kilometres above the ground. So, if rain is falling from a cloud but evaporating before it reaches the surface, rain may be reported when none is reported at the ground. Similarly, if rainfall is being enhanced near the surface (for example, by the seeder-feeder mechanism; see Chapter 13), low-level heavy rain may also be missed. Another problem to be accounted for is **attenuation**. Patches of heavy rain can remove so much of the initial pulse that very little of it passes into the area behind.

Modification of the wavelength of the pulse can also allow the radar to detect particles and droplets of different sizes. Most northern European rainfall radar uses microwaves with a wavelength of 56 mm, while the US network uses a wavelength of 100 mm – the latter suffers less from attenuation, but is less sensitive. Tuning the wavelength down to 30 mm allows the detection of cloud particles. Cloud monitoring by radar is not performed operationally, but ground-based observations of cloud provide valuable information for researchers. Exchanging the microwave pulse for a pulse of light (referred to as **lidar**, from 'light detection and ranging') allows the detection of even smaller targets, such as aerosols, small insects and some larger molecules. Lidaar can also be used to detect the height of the cloud base, although it struggles to detect anything within a cloud, as the cloud reflects light so strongly. For safety, lidar tends to use ultraviolet radiation at a wavelength of 350 nm as opposed to visible light.

A raw radar return contains data at a range of different horizontal resolutions. Near the radar, the beam is much narrower and can therefore resolve much more detail than it can at its maximum range. It also contains overlapping data from each of the radar sites. The data is therefore collected and mapped onto a grid by computer. This process also removes all the spurious returns from solid objects, mirages and so on. The data is then ready to be fed into the forecast model – or used on a TV forecast to indicate recent rainfall patterns.

Another advance in radar technology is the use of **Doppler radar**. This can be used to provide information about the patterns of wind within areas of rainfall, giving the capability for warnings of developing tornadoes and downbursts, the latter of which is particularly hazardous to aviation. Doppler radar measures the time separation of the pulses, which increases if the rain particles are moving away from the transmitter or decreases if the rain is moving towards the transmitter. As the Doppler shifts are usually very small, the time change is determined in terms of a shift in phase of the radio waves. Since the early 1990s, Doppler radars have been gradually replacing the traditional radar installations in many countries. In the UK, just over one-third of the radar network has Doppler capabilities; in the USA, the entire 159-strong **Nexrad** ('Next Generation Radar') network is made up of Doppler radars.

4.3 Observations from Space

The beginning of the satellite era was marked by the first artificial satellite launched into space, Russian satellite Sputnik 1 in 1957. Early results from this mission revealed that space provided an excellent platform from which to monitor the Earth's atmosphere and provide additional data to supplement the observations made on the surface. Early weather observation satellites carried downward-pointing cameras that took images of the Earth from above and beamed them back to Earth via radio. Since these early satellites, observation techniques from space have grown rapidly. Modern satellite pictures still consist of images of the clouds and weather systems surrounding the Earth, although cameras have now been replaced with radiometers that can sense radiation over a number of different wavelengths, enabling Earth observation even during times of darkness. Satellites allow real-time monitoring of the atmosphere: an image from a satellite can be instantly transmitted back to the ground.

A plethora of weather satellites orbit the Earth nowadays, launched and managed by many different countries, but all providing valuable data for both forecasting and research to the world's weather agencies and universities. Weather satellites fall into two main categories. First is the **geostationary satellite**, which orbits the Earth at a height of about 36,000 km, directly above the Equator. At this height, the orbital period of the satellite is exactly one day, allowing the satellite to view the same side of the Earth at all times. Satellites return pictures of the entire disc of the Earth that is facing them, with the best view of the tropics. Images of the weather at higher latitudes, however, tend to be of poorer quality on account of the Earth's curvature. Currently, a fleet of geostationary satellites, managed by agencies in the USA, Japan, India and Europe, monitors the weather worldwide.

A second (and much more numerous) fleet of satellites keeps track of the weather from **polar orbits** or **low-Earth orbits**. They travel around the Earth in a much closer orbit (heights are typically about 850 km up) and perform one orbit in about 90 to 100 minutes. They orbit over the poles, providing images as they travel northwards or southwards on their journeys. As they are in lower orbit, they provide much clearer images – particularly over higher latitudes, where geostationary satellite images are poor. Also, unlike their

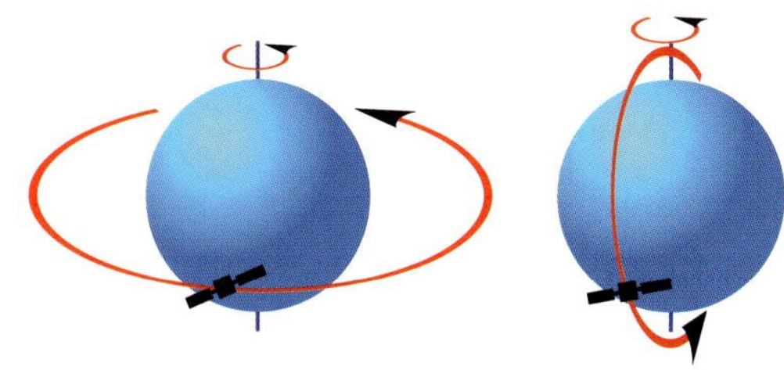

Figure 4.6 Polar and geostationary orbits around the Earth.

geostationary cousins, they can view the whole Earth; as they orbit, the Earth turns slowly beneath them, allowing their north–south paths to precess slowly to the west (Fig. 4.6).

The images they return are far improved on the early images provided by the cameras aboard weather satellites. On older satellites, radiometers built up images using three wavelength channels. Visible satellite pictures (Fig. 4.7A) show the Earth as an observer orbiting the Earth in a spacecraft would see it, detecting reflected radiation from the Sun. However, this only provides information about the clouds when the satellite is on the Sun-facing side of the Earth; in hours of darkness, visible satellite images appear black. Therefore, a second wavelength

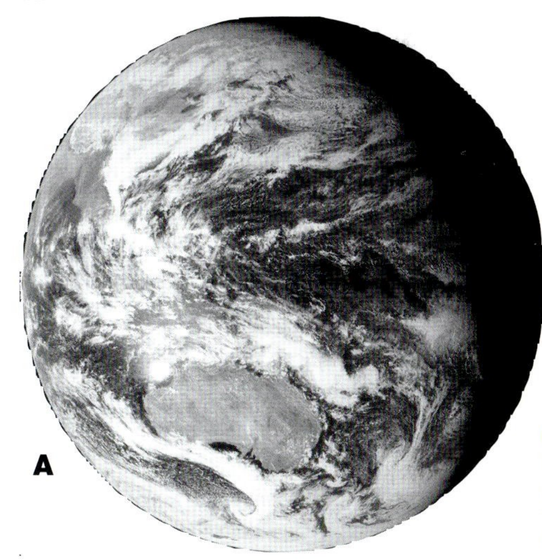

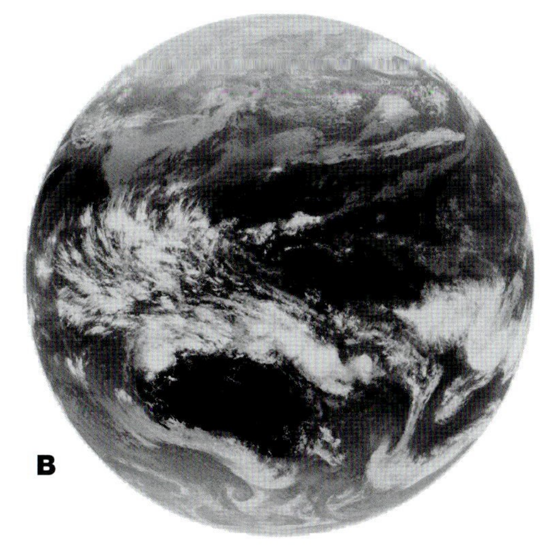

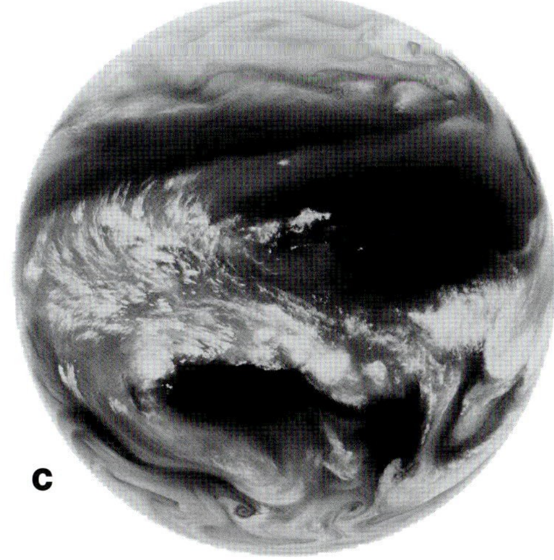

Figure 4.7 Geostationary satellite images of the western Pacific at 06:00 UTC on 23 January 2011 when viewed in (**A**) visible light, (**B**) thermal infrared and (**C**) the water vapour band (6.5 μm to 7.0 μm). (Images: NEODAAS/ University of Dundee.)

channel in the infrared is also used. Clouds emit radiation according to their temperature (see Chapter 7). An infrared image therefore has the advantage of not only detecting clouds at night, but also detecting cloud-top height (taller clouds have colder tops that emit radiation at lower intensities – Fig. 4.7B). A third channel often present is the so-called water vapour channel. This uses a wavelength at which water vapour emits radiation and allows detection of patterns of high and low humidity in the upper and mid-troposphere (Fig. 4.7C). It also allows winds to be determined in areas of clear sky.

Satellites have, of course, come far beyond just feeding cloud images back to Earth. Modern satellites include many more than these three channels, allowing even more information to be gleaned. Technology now allows space-based retrievals of all sorts of quantities. Profiles of temperature and humidity can be obtained by considering emission wavelengths of carbon dioxide; indications of wind speed and direction can be measured by cloud movements and even wave patterns on the surface of the ocean. Even from 850 km up, it is possible for a satellite to measure the height of the sea surface to an accuracy of the order of 2 cm. They can monitor the surface, both in terms of land use and snow cover, and track forest fires. They can also provide information about gas distributions in the atmosphere, such as **ozone**. Some new satellites even have downward-pointing radar and lidar on board to obtain profiles of the clouds from above.

5 Anatomy of the Atmosphere

The **atmosphere** is the place that weather calls home. Or perhaps it is more accurate to say that the atmosphere is weather – a complicated set of interactions between all of its constituent parts leading to wind, clouds, rain and weather systems. Weather is defined as the state of the atmosphere at a given time. So, if we are to understand the science of weather, it makes sense to start by becoming acquainted with the atmosphere (Fig. 5.1).

To create a weather forecast, a set of observations, however full, is not enough. We need to be able to describe and explain all the atmospheric processes that we refer to as weather, and how they all interact. Then, we need to convert this understanding to computer code that allows us to numerically model the weather. But that is a later stage: first, to begin our journey, we take a trip through the science of weather by examining the atmosphere itself.

5.1 Composition of the Atmosphere

Not only is the atmosphere home to all weather, it is also home to us. It provides us with the constant supply of oxygen that our bodies need to function – without it, there is no way that life on Earth could have survived. Removing this rich supply for just a few minutes can cause irreparable damage to

Figure 5.1 The Earth from aboard the International Space Station. The thin blue haze above the Earth's horizon is caused by light scattering off the atmosphere. (Photo: NASA Earth Observatory/ JSC Gateway to Astronaut Photography of Earth.)

our bodies, and possibly even death. We have evolved to survive only in the lower reaches of the atmosphere, where there is a sufficient quantity of **oxygen** – even only a few kilometres above us, the atmosphere becomes too thin to provide us with the air we need.

Fortunately, there is plenty of air to go around. The Earth's atmosphere contains a vast amount of air – about 5×10^{18} kg in total, with approximately 10,000 kg of air piled up on each square metre of the Earth's surface. The depth of the atmosphere is often taken to be about 100 km, although in practice it is hard to define exactly, as there is no distinct upper boundary – indeed, atmospheric molecules exist much higher than this. However, in comparison to the size of the Earth, whose radius is about 6,400 km, the atmosphere is only a very thin blanket covering its surface. In fact, if the Earth were to be scaled down to the size of an apple, the depth of the atmosphere would be comparable to the thickness of the apple's skin.

The Earth's atmosphere has certainly not always been as habitable as it is today. In the early days of the Earth's formation, the atmosphere was most likely a thin layer of hydrogen and helium – the most abundant chemical elements in the universe. At this time, the Earth would still have been a molten ball of hot rock spewing gases into the atmosphere, mainly in the form of **nitrogen**, **carbon dioxide** and **water vapour**. Gradually, the atmosphere would have filled with these gases, while the much lighter hydrogen and helium molecules would have escaped to space.

Over time, this carbon dioxide would have become trapped in rocks and the water vapour stored in the oceans, leaving nitrogen – an inert, unreactive gas – to eventually take over as the most abundant gas in the atmosphere. Over millions of years, primitive plant life would have begun converting some of the remaining carbon dioxide to oxygen. It is also possible that, at the same time, sunlight could also have been breaking down water vapour molecules into hydrogen and oxygen.

The present composition of the atmosphere is mostly molecular nitrogen (N_2) and oxygen (O_2). By volume, nitrogen accounts for 78% of the atmosphere and oxygen for 21%. Most of the remainder (nearly 1%) is **argon** (Ar). The rest is a mixture of trace gases at various concentrations, including carbon dioxide (CO_2), neon (Ne), helium (He), **methane** (CH_4), ozone (O_3), water vapour (H_2O) and xenon (Xe). Typical concentrations of these gases are given in Figure 5.2. Most of these gases are called **well-mixed gases** – they exist in the same percentages all the way up to 100 km above the surface. The exceptions are water vapour, which exists in much stronger concentration near the Earth's surface, and ozone, which is mostly found in a region commonly referred to as the ozone layer, between about 20 km and 30 km above the surface.

These gases are not confined to the atmosphere, but are part of a complex cycle that also affects the biosphere, the oceans and the geology of the Earth itself. For example, carbon dioxide is taken out of the atmosphere by **photosynthesis** in plants, and replaced by oxygen. This results in carbon storage within plant matter. The animal consumption of this vegetation converts it back to energy, and releases it back to the atmosphere by respiration. This is just a small component of the carbon cycle, which spans many different scientific disciplines – way beyond the meteorological limits of this book.

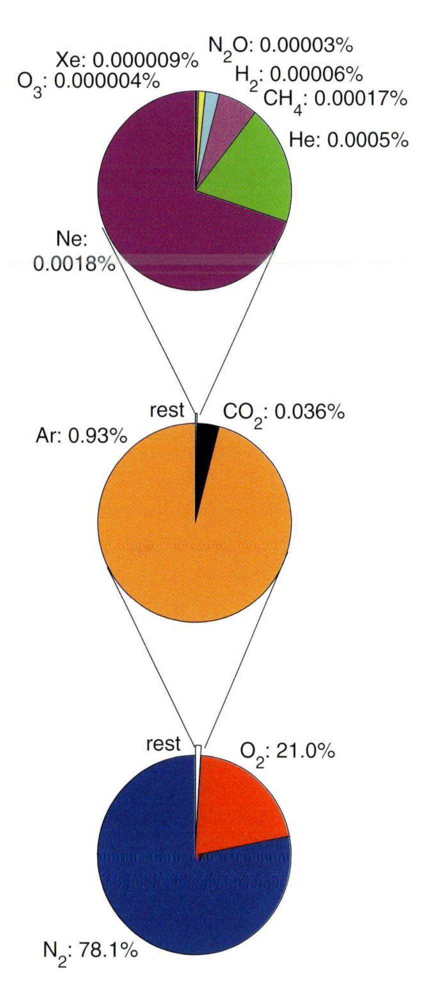

Figure 5.2 Composition by volume of dry air near the Earth's surface. Note that ozone concentrations are much higher than stated here within the ozone layer. Water vapour concentrations are highly variable: in very moist conditions, such as over ocean surfaces, air can consist of up to 4% water vapour (96% dry air).

5.2 Pressure, Temperature and Density

The weight of several thousand kilograms of air pressing down on us is what we experience as atmospheric pressure – although our bodies have evolved to not notice it. At the Earth's surface, the average atmospheric pressure we experience is 1,013.25 hPa. However, as we know from our experience of weather maps, atmospheric pressure at the surface can vary as weather systems pass over. Values of **mean-sea-level pressure** are generally found in the range 950 hPa to 1,050 hPa (although the centres of extreme low-pressure systems can fall below this range). Horizontal pressure gradients of 100 hPa over a few thousand kilometres are possible.

However, atmospheric pressure changes much more rapidly with height. A similar pressure change of 100 hPa would require a vertical journey of only 1 km. The pressure at a point in the atmosphere is dependent on the weight of air pressing down above that point. In other words, the pressure must decrease with height through the atmosphere. A barometer at a mean-sea-level pressure of 1,000 hPa would read about 940 hPa at the top of Kuala Lumpur's Petronas Towers, just under 900 hPa on the spire of Burj Khalifa in Dubai (the world's tallest building); and 350 hPa at the top of Mount Everest. This difference in horizontal and vertical pressure gradients shows how strongly stratified the Earth's atmosphere is.

Any climber taking on an ascent of Mount Everest will require oxygen supplies. This is because the **density** of the air is also much lower at this altitude. Density is defined as the mass of matter per unit volume, and the density of the atmosphere is inextricably linked to atmospheric pressure. Near the Earth's surface, the atmosphere typically has a density in the order of 1.2 kg m^{-3}. However, at the top of Everest, the air density has fallen to about 0.5 kg m^{-3} – in other words, less than half that at the surface.

In fact, both air pressure and density reduce exponentially with height above the Earth's surface. To explain the link between these quantities, we must consider the gases of the atmosphere on a molecular scale. It consists of a jumble of gas molecules, all moving around in random directions and bouncing off each other. The collision of a molecule with any object – be it another molecule or a solid object – results in a tiny force being exerted. We are being bombarded by molecules in this way all the time, and it is the sum total of all of these tiny forces that we experience as atmospheric pressure.

Despite their apparent random motions, the molecules of the atmosphere are constrained by **gravity**. Without gravity, the Earth's atmosphere would simply drift away into space. The pull of gravity causes the molecules of the atmosphere to gather in much greater numbers near the Earth's surface. Near the surface, the increased weight of the air above forces the molecules closer together, increasing both the pressure and the density.

The inextricable link between pressure and density can be explained by considering a solid, sealed box containing a fixed number of gas molecules (Fig. 5.3B). As they bounce around the box, they exert an outward pressure on the sides of the box as their random paths cause them to collide with it. We can increase the pressure within the box in two ways: we can either push one wall of the box inwards, therefore reducing the volume of the box, or we can pump in more gas (Fig. 5.3A). Both processes have the effect of increasing the density of gas within the box, and therefore increasing the frequency of collisions of molecules with the sides of the box. If we assume the walls of the box do not allow any heat to escape, density and pressure are proportional.

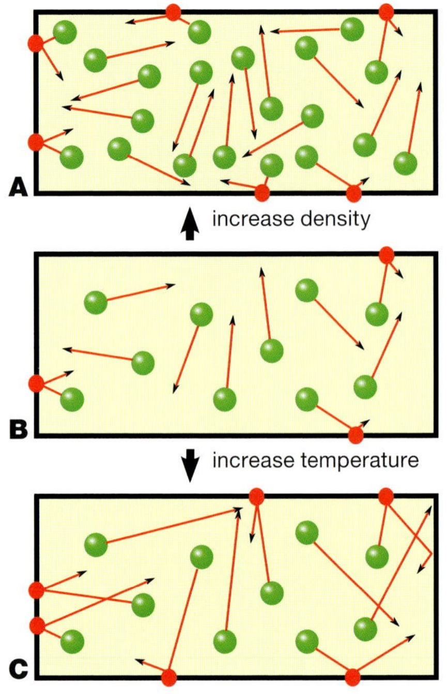

Figure 5.3 A collision of a molecule with the side of the box, here marked by a red spot, exerts a small amount of pressure. The sum total of these collisions is the pressure exerted by the molecules on the sides of the box (**B**). Both increasing the density by doubling the number of molecules (**A**) and increasing the temperature, which makes the molecules move faster (**C**), results in more of these collisions, hence an increase in pressure.

A third quantity plays a part in the relationship between pressure and density: **temperature**. On a molecular level, temperature is a measure of how much energy the molecules have; in other words, how fast they are travelling along their paths. If we maintain the same size box and keep the same number of molecules inside, but heat the air up, the molecules inside will therefore move more energetically. This causes an increase in pressure, as collisions between molecules and the side of the box become more frequent, as they are moving more quickly (Fig. 5.3C). The behaviour of the molecules in the atmosphere is the same, although of course the pressure of the molecules colliding with the walls of the box is now applied to the surrounding molecules.

5.3 The Atmospheric Profile

While the pressure and density profiles of the atmosphere are smooth and roughly exponential, the profile of atmospheric temperature is very different. The pressure and density profiles are driven primarily by the weight of the atmosphere itself. The temperature profile, however, is driven by the transfer of heat within the Earth–atmosphere system. This results in the heating of the atmosphere at different levels, creating pronounced temperature maxima and minima and dividing the atmosphere into a number of layers, as depicted in Figure 5.4. As the trend of temperature with height varies, the layers can have very different properties. We will return to the details of where the heat energy comes from in Chapter 7.

The lowest layer of the atmosphere, extending from the Earth's surface up to a height of 10 km to 15 km, is referred to as the **troposphere**. To the meteorological world, the troposphere is by far the most interesting and important of the layers. Not only is it the layer in which we live, but also the layer in which the majority of the weather phenomena mentioned in this book exist. Perhaps this is because it contains four-fifths of the mass of gas in the entire atmosphere.

Most of the heat supplied to the atmosphere comes from the Earth's surface. For this reason, the troposphere is warmest near

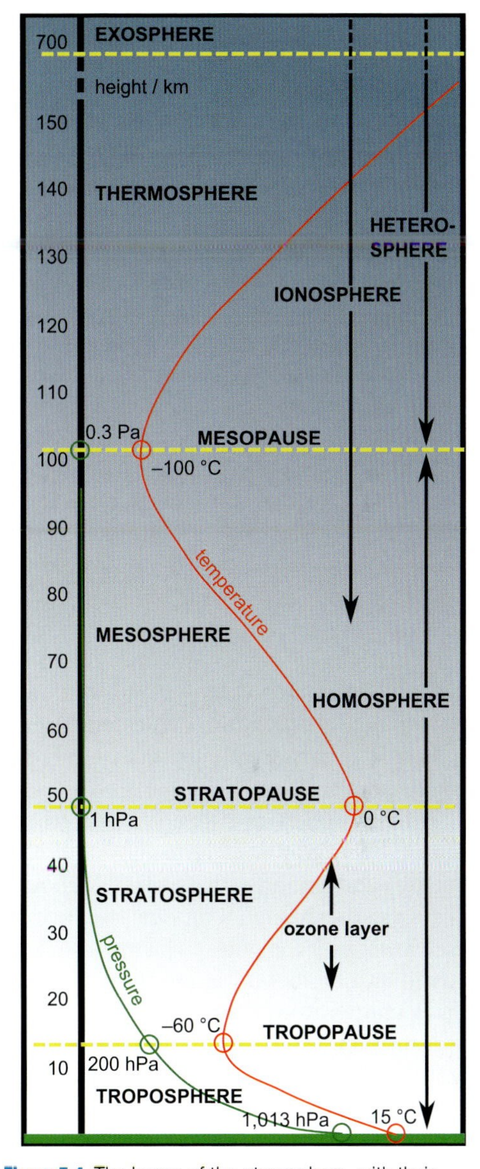

Figure 5.4 The layers of the atmosphere, with their approximate depths (note exact depths vary with season and latitude). Profiles of pressure (green) and temperature (red) are shown, with typical values at the tropopause, stratopause and mesopause indicated.

the surface, and becomes gradually colder with height. Also, the water vapour supplied to the atmosphere comes from the surface by evaporation and transpiration. On account of this, the troposphere contains nearly all of the water vapour in the entire atmosphere. Heat and moisture are transferred from the surface to the troposphere via a turbulent layer right at the bottom, often referred to as the **boundary layer**.

The flow of air in the troposphere is very disturbed and turbulent, and characterised by eddies and vortices. Such atmospheric circulations are observed on scales of thousands of kilometres down to very small local circulations a few centimetres across. These turbulent, overturning circulations transport both heat and water vapour around the atmosphere and drive the Earth's weather machine – the circulations themselves create winds and differences in pressure; the moisture carried upwards creates clouds and rain. The next few chapters are dedicated entirely to these weather-creating processes in the troposphere.

The troposphere is capped by the **tropopause**. The tropopause is usually found at 10–15 km above the Earth's surface (although it can get much higher in the tropics and much lower over the poles). The tropopause is marked by a temperature minimum (referred to as a **temperature inversion**) and indicates the transition layer between the troposphere and the **stratosphere**. As a temperature inversion acts as a cap for circulations, the **turbulent flow** of the troposphere is incapable of penetrating far into the stratosphere. Hence, the movement of air in the stratosphere is not turbulent, as it is in the troposphere, but **laminar** (flowing in layers). As there is so little transport of air across the tropopause, the air in the stratosphere contains very little moisture and is largely clear of clouds.

Throughout the stratosphere, the temperature increases with height. The cause of this behaviour is the fact that the stratosphere contains the **ozone layer**. Ozone absorbs ultraviolet radiation from the Sun via a series of chemical reactions, heating the layer. Maximum heating occurs towards the top of the layer, creating a temperature maximum at a height of about 50 km above the surface (referred to as the **stratopause**).

By the time we have reached the stratopause, we have already accounted for over 99.9% of the atmosphere's gas molecules. The air pressure here is 1 hPa and the typical density of the air is about 0.001 kg m^{-3}. At this point, we have reached the top of the layer of stratospheric ozone; hence the warming effect is lost above the stratosphere. As a result, the layer above, referred to as the **mesosphere**, once more becomes cooler with increasing height. Turbulent flow can occur in the mesosphere, as in the troposphere, and sometimes **noctilucent clouds** can form over the poles. These are very thin, tenuous ice clouds found in the middle of the mesosphere (Fig. 5.5). Their origin is uncertain, although theory suggests that the water vapour used to condense them originates from the upper stratosphere and lower mesosphere, and is carried upwards by the mesospheric circulation. Burnt-up remnants of meteors act as nuclei on which the clouds can form. Circulations in the mesosphere cool the air as it ascends, leading to the top of the mesosphere being the coldest naturally occurring part of the Earth–atmosphere system, with temperatures typically around –100°C.

Figure 5.5 Noctilucent clouds over Reading, UK, on 14 July 2009. These clouds consist entirely of ice and form in the mesosphere. As they are so high up, they are often illuminated by the Sun after sunset. (Photo: Stuart Nock.)

The top of the mesosphere varies in height from 80 km to 120 km. Above this layer we find the **thermosphere**, where temperature again increases with height. This heating is caused by direct interaction of air molecules with energetic particles from the Sun, heating the air up to temperatures of about 1,500°C. However, the density of the thermosphere is so low that an object travelling in the thermosphere could never be heated to this temperature, as there would be too few molecular collisions to transfer enough heat energy. The interaction of solar rays with gas molecules also leads to the process of **ionisation**, where molecules are stripped of electrons and become charged. This creates a charged layer high up in the atmosphere, often called the **ionosphere**.

Beyond about 100 km from the Earth's surface, the well-mixed nature of the atmosphere starts to break down, with the gases of the atmosphere sorting themselves into layers according to their molecular mass. This stratified upper portion of the atmosphere is sometimes referred to as the **heterosphere**, in contrast to the well-mixed **homosphere** below. The heterosphere extends hundreds of kilometres into space.

6 Water in the Atmosphere

We now return to the troposphere, where the atmosphere is dense and the water vapour is plentiful. Of all of the gases present in the troposphere, water vapour is certainly the one that has the largest impact on the weather we experience. The difference between a fine weather day and a poor weather day is often the presence of clouds and precipitation – both of which develop from the condensation of water vapour. Perhaps the most interesting aspect of water in the atmosphere is that, in the range of temperatures that occur in the troposphere, it can exist in all three phases (gas, liquid and solid). Precipitation can be both liquid (rain) and solid (snow, hail and graupel) in nature, as indeed can cloud particles. The interaction of water in these different phases is an important step in understanding atmospheric processes.

6.1 The Hydrological Cycle

Along with the oxygen in the atmosphere, water is another of the fundamental requirements for our existence. Both plants and animals require water to survive – without water, we would not last long. Fortunately, water is plentiful on Earth in a number of different forms – about 1×10^{21} kg of water are contained in the Earth–atmosphere system. Nearly all of this (97% or so) is stored in oceans. Most of the remaining 3% is either locked in glaciers or polar ice caps, or stored within the Earth as groundwater. A very tiny fraction of the Earth's water – about 0.001% – is suspended in the atmosphere.

Water flows freely between these massive reservoirs. It evaporates off the surface of the oceans, rivers, lakes and puddles and enters the air in the form of water vapour. Aloft, it

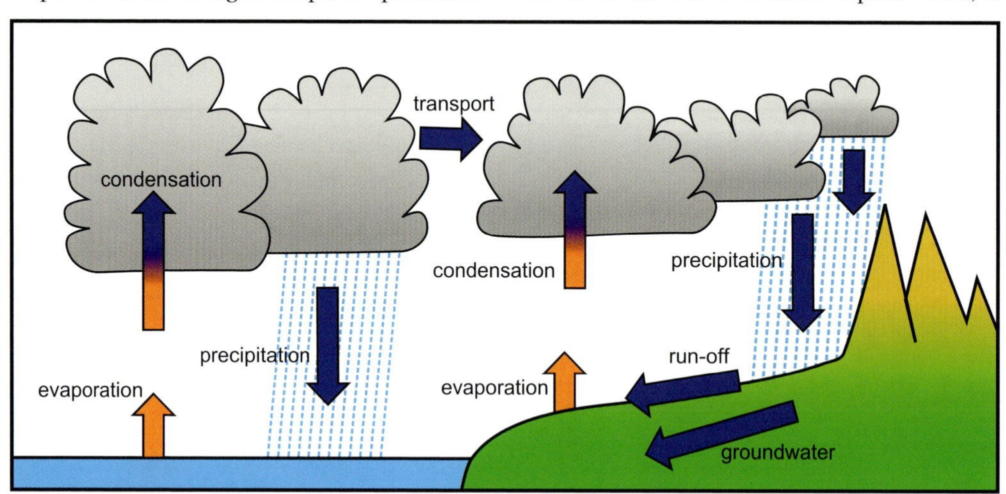

Figure 6.1 The hydrological cycle.

condenses to form clouds. If these become thick enough, they start to produce raindrops, hailstones and snowflakes, which fall to the ground as precipitation – either back into the ocean, or on land. In the latter case, the water soaks into the soil, gradually working its way via groundwater back to the river systems, again ending up in the sea. There are many branches of Earth sciences studying various elements of this **hydrological cycle** (Fig. 6.1). To understand the weather, the part of the cycle of most interest is that part where the water is aloft in the atmosphere.

Water molecules enter the atmosphere from the surface by evaporation. **Evaporation** is best explained on a molecular level, in a similar way to the atmosphere described in Chapter 5. Consider a puddle of water, as shown in Figure 6.2. On a molecular level, a gas is a large number of molecules all moving about in random directions. In a liquid the same is true, but with stronger bindings between the molecules. So molecules within the puddle of water can still move around and, if energetic enough, they can break free of its surface and be carried away into the atmosphere as water vapour. Atmospheric water vapour molecules bounce around among the air molecules, and some will be returned to the puddle. However, as long as there is a net flow of water molecules out of the puddle, evaporation will take place spontaneously.

Several factors affect the rate at which water can evaporate off the surface. If the puddle is warmer, the molecules in the puddle will be more mobile and therefore more likely to escape the surface. Also, if a breeze blows over the surface of the puddle, it will evaporate more readily. In the absence of a breeze, the rate of evaporation will gradually slow down, as drier air can receive more water vapour than air that is already humid. A breeze provides a constant supply of drier air into which the puddle may evaporate.

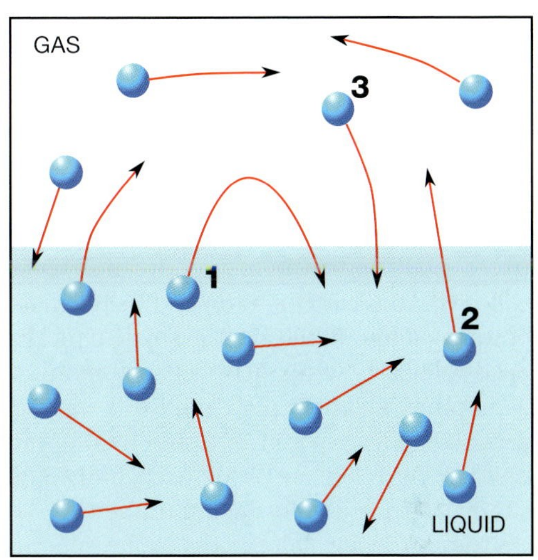

Figure 6.2 The process of evaporation on a molecular scale. Some molecules within the liquid have sufficient energy to break free of the surface. Some of these fall back into the liquid (1), while others escape into the air above (2). At the same time, water molecules in the air may pass back into the liquid (3). Evaporation will only take place if more molecules leave the liquid than return to it.

From the surface, the water vapour can then be carried up into the troposphere. Because it enters the atmosphere from the surface, there tends to be much more water vapour in the lower troposphere than higher up near the tropopause. The fact that upper tropospheric air is so dry means that only very small amounts of water vapour make it into the stratosphere and the layers above.

6.2 Humidity and Moisture

The moisture content of a mass of air is referred to as its humidity, and there are several ways of expressing the amount of water vapour in the air. **Specific humidity** describes the mass of water vapour per unit mass of air, expressed in units of $kg\,kg^{-1}$. Typical values of specific humidity in the lower troposphere are of the order of $0.01\,kg\,kg^{-1}$ (or $10\,g\,kg^{-1}$). Alternatively, humidity can be expressed as the mass of water vapour per unit mass of dry air. This is called the **mass mixing ratio** and, as the mass of water vapour in the air is so small, typically has a value very similar to specific humidity.

Another measure of humidity is **vapour pressure**. Air pressure is the sum total of the pressures exerted on us by each individual gas species present in the atmosphere. So, if we were to remove all the gas in the column of air above us except, for example, oxygen, we would experience the **partial pressure** that is due to the contribution of oxygen alone – which would be less than the total atmospheric pressure. Vapour pressure is therefore the contribution to total atmospheric pressure that is exerted by the minuscule amounts of water vapour that exist in the atmosphere. Typical values are 15 hPa (about 1.5% of the total atmospheric pressure).

Whatever method we use to quantify humidity, its value increases as our puddle evaporates into the air above it. However, as the humidity of the air increases, the flow of water vapour from the air back into the puddle becomes greater, implying that the air can receive less and less extra water vapour from the surface of the puddle. This could, in theory, continue until the flux of water vapour into and out of the puddle becomes constant. In this instance, the air above the puddle can hold no more water vapour and reaches saturation. However, the saturation point in terms of specific humidity, mass mixing ratio and vapour pressure is dependent on the local conditions of temperature and pressure. The definition of **relative humidity** makes the situation a great deal simpler. Relative humidity is the quantity of water vapour contained in the air as a percentage of the quantity it could hold if it were saturated. Hence, saturated air always has a relative humidity of 100%; dry air a relative humidity of 0%.

6.3 Water Droplets and Rain

Water vapour on its own, though, does not directly form clouds or precipitation. To form a cloud, this water vapour must be converted back into either liquid water or solid ice. But how do we get the vapour to condense back out? The answer lies in the definition of saturation. The maximum amount of water vapour that a mass of air can hold is related to the temperature of the air, with warmer air capable of holding more water vapour than cold air. In other words, by raising or lowering the temperature of the air, we can change the maximum amount of water that the air can hold. So, if we cool down a mass of moist air but do not allow water vapour to leave, the air's relative humidity will climb until it reaches a temperature at which it is saturated. This temperature is referred to as the **dewpoint temperature**. Further cooling of the air results in **supersaturation** and, if the conditions are right, condensation will take place and the beginnings of a liquid cloud will form.

In supersaturated conditions, there are two methods by which water droplets may form. The first is by a process known as **homogeneous nucleation**, in which water droplets condense directly out of the air. However, this is not an efficient method for droplet formation; droplets that form by this method are so small that they can quickly re-evaporate, even under supersaturated conditions. In their short lifetime, there is not enough time for the droplet to coalesce with other droplets and start growing. In fact, under homogeneous nucleation, droplets only start to form in any numbers in extremely supersaturated air – with a relative humidity of 300% to 400%. Such high levels of supersaturation never occur in the atmosphere; relative humidity with respect to water rarely exceeds 101%.

The second method of droplet formation – the process by which most droplets form in the atmosphere – is **heterogeneous nucleation**. We often think of the atmosphere as a pristine layer consisting solely of a mixture of gases. But, there are also a very large number of tiny, solid particles that are light enough to move around the atmosphere with the wind. These are referred to as **aerosols**, examples of which include soot from forest fires, dust from deserts and salt particles from the sea. Aerosols act as nuclei onto which condensation can more readily occur. They are also useful in preventing re-evaporation, as a droplet containing an aerosol particle presents less surface area of water through which evaporation can occur than a pure water droplet (Fig. 6.3).

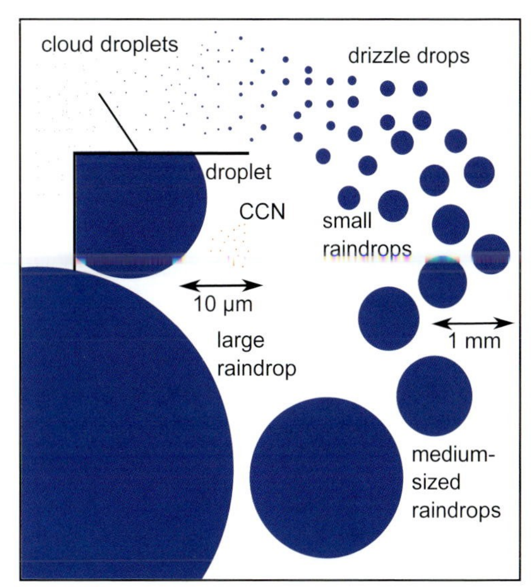

Figure 6.3 Typical sizes of raindrops, cloud droplets and cloud condensation nuclei (CCN), all shown to scale. The smallest cloud droplets are about 1 µm across. Drizzle drops are in the range 100 to 500 µm across; large raindrops can reach up to 5 mm across.

Once a droplet has formed it has the potential to grow. Growth of a droplet occurs by two processes. Firstly, as droplets are carried around within a cloud, they can collide with one another and merge to form larger droplets. This is called **coalescence**, but is not a particularly effective method of growing water droplets. Typical densities of liquid water in a cloud are $1\times10^{-4}\,kg\,m^{-3}$ – in other words, droplet collisions do not happen frequently. A more efficient process of droplet growth is simply by condensation. Where a droplet forms, there will be a reduced amount of water vapour surrounding it, as this vapour has been used to form the droplet. In nature, a fluid will move from a high concentration to a low concentration by **diffusion**. This occurs around the water droplet, providing it with a flow of fresh water vapour to condense onto its surface.

If a droplet grows large enough, it will, of course, eventually fall out of the cloud as rain as it becomes too heavy to be supported by the circulation within. The typical size of a newly formed droplet is about 1 µm (this is the rough size of the smallest possible aerosol particle on to which condensation occurs). Raindrops can take a wide range of sizes. Drizzle consists of a large number of very small raindrops about 200 µm across, but heavy showers can generate raindrops that are much larger – up to a few millimetres. The maximum size that a raindrop can achieve is largely dependent on the time it has spent in a cloud, and hence the time it has had to grow. The largest raindrops rarely exceed 5 mm in size: raindrops larger than this tend not to have sufficient surface tension to hold them together. Contrary to expectation, raindrops are not shaped like the teardrops that often represent rain on weather forecast maps. Smaller raindrops are almost perfect spheres; larger ones take the shape of flattened (**oblate**) spheres.

6.4 Ice Crystals and Snowflakes

Temperatures in the troposphere can fall a long way below the freezing point of water. Hence it is not surprising that water vapour can also condense in the form of ice crystals. In fact, at any one time, a great deal of the troposphere by volume is below 0°C. The height at which the temperature has fallen to 0°C, referred to as the **melting layer**, or freezing layer, is usually 2 to 3 km up in the mid-latitudes and 4 to 5 km in the tropics. This suggests that a high percentage of all condensation that occurs in the atmosphere should be in the form of ice particles. However, this is not the case: liquid water droplets can still form in great numbers at temperatures way below the freezing point of water. Such droplets are referred to as **supercooled liquid water** droplets. They can exist in the atmosphere at temperatures all the way down to –40°C, and it is from these supercooled liquid water droplets that ice crystals can form.

For one of these supercooled droplets to become ice, some sort of ice nucleation event must occur within it. When some of the molecules in a supercooled droplet group together to form an ice nucleus, the entirety of the drop will quickly freeze and form an ice crystal. This nucleation process occurs more readily at lower temperatures, when the water molecules in the droplet are moving around less energetically and are more likely to align themselves in the formation of an ice crystal. Below –40°C, this is the dominant process by which ice crystals form, and is referred to as homogeneous nucleation in analogy to its liquid equivalent: the ice crystal forms from the drop with no outside contribution. At higher temperatures, heterogeneous nucleation can also occur. As for the liquid water equivalent, this process requires an aerosol nucleus to initiate the rapid freezing of a supercooled droplet.

The presence of ice further complicates the concept of humidity, too: a mass of air with a given relative humidity with respect to water can have a very different relative humidity with respect to ice. In a supercooled liquid cloud, the relative humidity with respect to water may be 100% but the relative humidity with respect to ice can be very much larger – sometimes over 150%. In other words, air within a supercooled cloud can be saturated with water, but extremely supersaturated with respect to ice. This is because, at a given temperature and pressure, water vapour evaporates off the surface of a liquid droplet far more readily than it does off the surface of an ice crystal. Hence, the air must hold a great deal more water vapour to be saturated with respect to a liquid droplet than it needs to be saturated with respect to an ice particle. This means that, if ice nuclei are inserted into the supercooled cloud, ice crystals will instantly start to form in great numbers at the expense of the liquid cloud. Depicted in Figure 6.4, this is referred to as the **Bergeron–Findeisen**

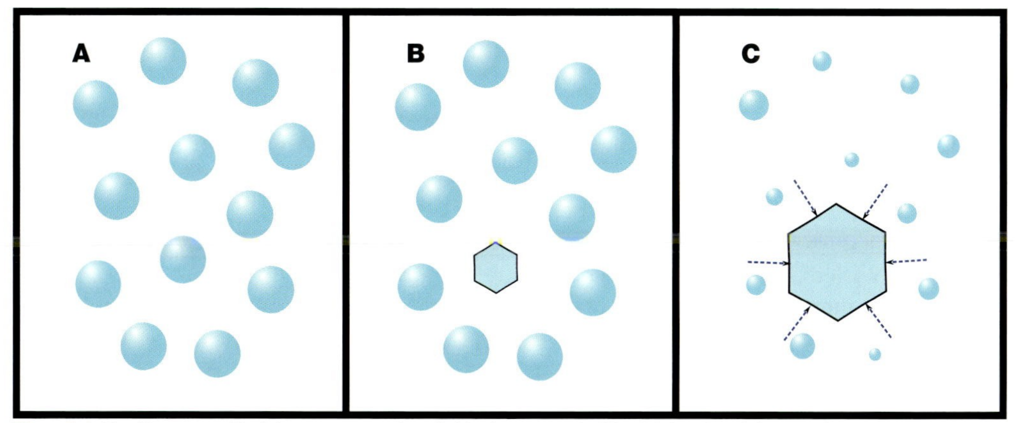

Figure 6.4 The Bergeron–Findeisen process. In a field of supercooled liquid droplets (**A**), one of the droplets may freeze to form an ice crystal (**B**). This results in the rapid growth of the ice crystal at the expense of the liquid droplets (**C**).

Figure 6.5 A hole-punch cloud over Park City, Utah, USA. Each hole was generated by the passage of an aircraft through the altocumulus layer. (Photo: Mike Blackburn.)

process, which can sometimes be seen in action when an aircraft flies through a layer of supercooled cloud. This can cause ice nucleation, initiating rapid glaciation of an area of the cloud, resulting in a circular hole in the cloud and a stream of ice crystals falling beneath. This phenomenon is often seen near airports, and is called a **hole-punch cloud** (Fig. 6.5).

Another major difference between the formation of water droplets and ice crystals is their shape. Liquid droplets are limited to being spherical (or slightly oblate if they are falling as rain). Ice crystals can, however, take on a plethora of different shapes depending on the conditions in which they form. Ice tends to form in a hexagonal lattice, taking on a shape with some form of six-fold symmetry. The smallest crystals tend to be either flat hexagonal plates or long, thin rods that are hexagonal in cross-section. As they grow, they can develop many shapes, often culminating in either snowflakes (Fig. 6.6) or random clusters of ice crystals, all stuck together if the temperature is cold enough. In warmer conditions, ice crystals can sweep out and accumulate liquid water droplets to form a lump of snow, referred to as **graupel**.

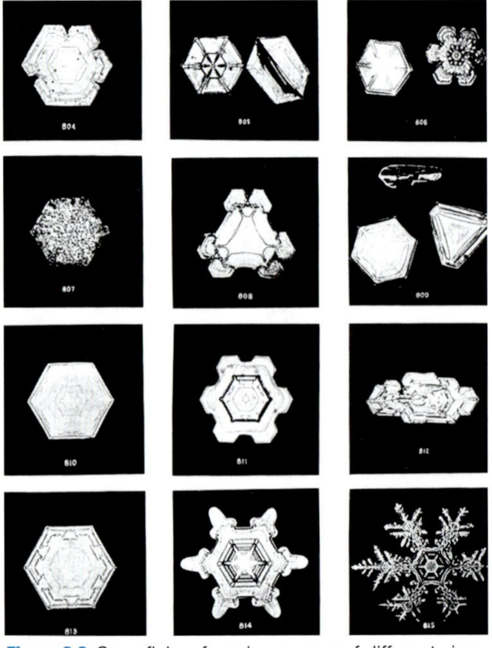

Figure 6.6 Snowflakes form in an array of different sizes and shapes. (Image: NOAA Picture Library.)

When the ice crystals become heavy enough to overcome the circulation within their parent cloud, as for liquid droplets, they start to fall. If the air is cold enough all the way to the ground, as may be the case in winter, they can reach the ground in the form of snowflakes or ice pellets. Partially melted snow is often referred to as sleet. In the summer, however, the snow will melt to raindrops at the freezing layer. Even in the summer, a large amount of the raindrops that fall started their journey to the ground as ice crystals or snowflakes.

7 It All Starts with the Sun

Having introduced the atmosphere, we now need to set it in motion. Evaporation of water off the surface may happen spontaneously, but for clouds to form, we need moist air to be cooled. For high-pressure and low-pressure weather systems to develop and winds to blow, we need both horizontal and vertical movement of air. In short, to make any weather happen, we need to provide energy to the atmosphere. However, this is not necessarily an easy task – recall that the atmosphere contains approximately 5×10^{18} kg of air, and keeping such a vast volume of air in motion requires a vast amount of energy. Fortunately, there is an energy source that is strong enough to handle this task – it sits at the heart of the Solar System.

7.1 Our Local Star

Along with the air that we breathe and the water that we need to remain hydrated, heat energy from the Sun is another vital component to life on Earth. Not only does it provide the energy required to drive the world's weather systems, it also provides us with both light and heat. Without the Sun's energy, the Earth would be a dark, freezing cold ball of rock in space, heated only from within by its own core – far too cold to sustain life.

The Sun, one of billions of stars in our local galaxy, is termed by astronomers a **yellow dwarf star**. As stars go, it is fairly small but, with respect to the size of the Earth, its dwarf status is perhaps misleading – the solar diameter is about 1,400,000 km, and its mass is about 2×10^{30} kg. In contrast, the Earth has a diameter of 12,800 km and a mass of 6×10^{24} kg: the volume of the Sun is over one million times that of the Earth. The Sun's core is a massive nuclear reactor running at a temperature of some 15,000,000°C and a pressure 250 billion times that at the surface of the Earth. In such extreme conditions, matter exists as a dense mass of hydrogen and helium nuclei. The combination of hydrogen nuclei and free electrons produce helium nuclei and massive amounts of energy, which streams out through the body of the Sun. The outermost layer of the Sun is called the **photosphere** and, while much cooler than the centre of the Sun, at about 5,500°C is still hotter than any temperature naturally occurring on the Earth's surface. It takes energy 100,000 years to travel from the Sun's centre to its surface.

It is the energy released from this outermost layer of the Sun (a total power output of 4×10^{26} W) that powers the atmosphere. This energy reaches the Earth as a combination of visible light, **ultraviolet** and **infrared radiation** – all forms of electromagnetic waves. They travel at a universally constant speed of 300,000 km per second, but have a wide range of wavelengths and frequencies and hence very different properties (Fig. 7.1). The longest wavelengths belong to radio waves, which can reach thousands of kilometres; X-rays and gamma rays lie at the other end of the spectrum, with wavelengths of the

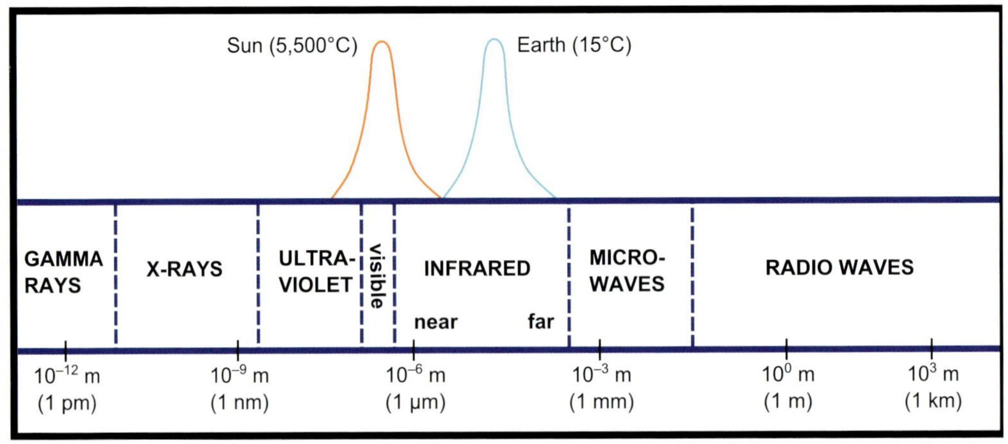

Figure 7.1 The electromagnetic spectrum. The different types of electromagnetic radiation are indicated with their typical wavelength ranges. The wavelengths over which the Sun and the Earth emit radiation are indicated by the orange and blue curves at the top.

tiniest fraction of a millimetre. The shortest gamma rays can have wavelengths of a picometre (10^{-12} m). Visible light has wavelengths in the range of 400 nanometres (400×10^{-9} m; blue light) to 700 nanometres (700×10^{-9} m; red light). The Sun emits most strongly at about 500 nanometres, which is in the yellow part of the spectrum.

7.2 The Earth in Equilibrium

Everything emits **electromagnetic radiation** – the Sun, the Earth, even ourselves. In fact, everything that has a temperature greater than absolute zero (the lowest theoretically possible temperature: 0 K or –273.15°C) must emit electromagnetic radiation at some wavelength and some intensity. Hotter objects emit higher energy radiation, which corresponds to shorter wavelengths. As they contain more energy, hotter objects also emit greater amounts of radiation. Temperatures at the surface of the Sun are about 5,500°C; a typical Earth surface temperature is about 15°C. Clearly, the nature of the radiation emitted by the Sun and the Earth must be very different.

The ranges of wavelengths over which the Sun and the Earth emit electromagnetic radiation are shown on Figure 7.1. In this figure, we make the assumption that the Sun and the Earth emit radiation at maximum efficiency at all wavelengths (**black bodies**). The wavelengths of the radiation they emit overlap each other very little. The peak radiation wavelength emitted by the Earth is in the near infrared, at about 10 μm. Radiation from the Sun is often referred to as **solar radiation** and mostly consists of light; radiation from the Earth is referred to as **thermal radiation** (or **terrestrial radiation**) and is mostly heat.

In other words, while the Sun warms up the Earth–atmosphere system, the Earth emits radiation to cool the system down. The balance of incoming solar radiation and outgoing thermal radiation is the Earth's **radiation budget**. Averaged over long periods of time, the sum of the incoming and outgoing

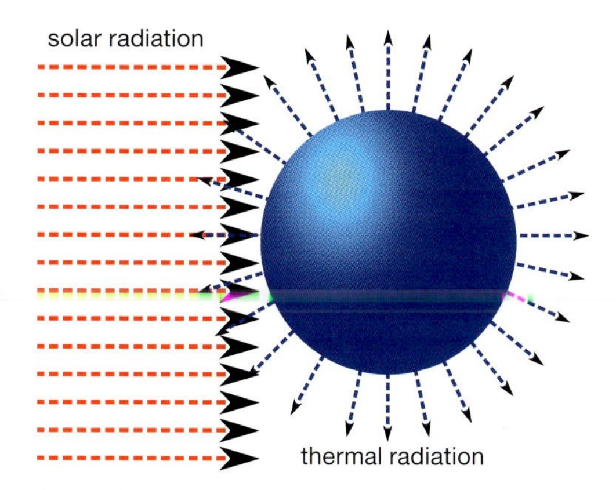

Figure 7.2 Averaged over long periods of time, incoming radiation from the Sun (to the left) is balanced by the thermal radiation emitted in all directions by the Earth.

radiation must be zero. If this were not the case, the Earth would either warm or cool. A system in balance in this way is said to be in **equilibrium** (Fig. 7.2).

The Earth receives an input of energy of about 1,360 W m^{-2} from the Sun – a value often referred to as the **total solar irradiance**. To heat up the Earth, this energy must be absorbed. However, only a fraction is actually absorbed by the Earth's surface; the rest is reflected back to space. The reflectivity of the Earth is referred to as its **albedo**: a surface with an albedo of zero absorbs all the radiation that falls on it; a surface with an albedo of 1.0 reflects it all. The mean albedo of the Earth's surface is approximately 0.15, although different surfaces can have very different values. Soil, forests and vegetation are typically much less reflective, with albedo values as low as 0.05. In contrast, desert surfaces typically have higher albedos, more in the range 0.3 to 0.4, while snow has a much higher albedo on account of the fact that it is white (up to 0.8 for pristine snow).

The absorbed fraction of this energy heats up the Earth's surface and can then be emitted as thermal radiation. But again, the emission of radiation from the Earth's surface is not completely efficient. The 'black-body' curves shown in Figure 7.1 assume that both the Sun and the Earth emit as much radiation as they can at all wavelengths. In reality, only a fraction of this maximum possible emission is actually emitted. **Emissivity** is a measure of how efficiently an object emits radiation. It turns out, however, that representing the Earth as a perfect emitter is not a bad assumption: most of its surface has values of emissivity in the range 0.90 to 0.98. Water is an almost perfect emitter; hence emissivity of the oceans is very near to 1.0.

7.3 The Effect of the Atmosphere

The atmosphere has a massive impact on the Earth's radiation budget. It may appear largely transparent to us but, in terms of thermal radiation, at some wavelengths it is virtually opaque and strongly absorbs energy. It also emits radiation: we mentioned previously that all objects must emit electromagnetic radiation – this includes the atmosphere and all its constituents. Additionally, it can also redirect radiation by the process of **scattering**.

To understand the interactions between radiation and the atmosphere, we must return once more to our molecular model of the atmosphere. At this scale, incident radiation takes the form of **photons**, small packets of energy travelling at the speed of light through the jumble of randomly moving molecules. Figure 7.3 shows the passage of photons through gas molecules. If a photon meets a molecule, two things can happen. The molecule can absorb the energy of the photon,

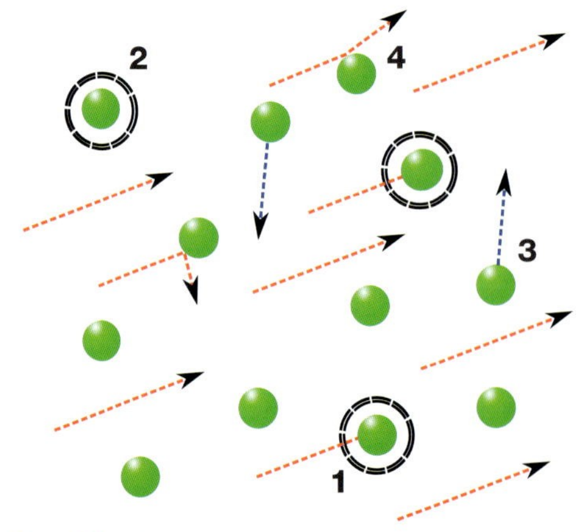

Figure 7.3 Photons (arrows) passing through a field of gas molecules. Some photons meet molecules and are absorbed (1), exciting the molecule. The molecule remains in an excited state (2) until it emits a photon – not necessarily at the same wavelength as the absorbed one (3). Some photons are also redirected or scattered when they meet molecules (4).

giving the molecule more energy and enabling it to move around faster. Alternatively, the photon can be deflected, or scattered, into a different direction. At any point, an energetic molecule can also emit a photon, which reduces its energy and slows it down.

The interaction between molecules and radiation is complicated by the fact that the atmosphere is not made entirely of a single gas. The interactions of each gas with both solar and thermal radiation are different, as the molecules contain different numbers of atoms and have different masses. In the atmosphere, solar radiation is often partitioned into direct and diffuse components. Direct solar radiation is the component that reaches the surface without interaction with the atmosphere; diffuse radiation has undergone at least one scattering event during its passage. So we only see **direct radiation** coming straight from the Sun, but **diffuse radiation** reaches us from all parts of the sky. Nitrogen and oxygen molecules, which together make up 99% of the atmosphere, are strong diffusers of radiation by scattering. They scatter sunlight equally in all directions, with shorter wavelengths of radiation (the blue end of the visible spectrum) scattered more effectively. For this reason, on a clear day, all parts of the sky appear blue. The same effect also explains the reds and oranges we see at sunset: the path of the sunlight through the atmosphere becomes much greater so, by the time the light reaches us, all the blue light has been scattered out, and the sunset appears red (Fig. 7.4).

Vast quantities of incoming solar radiation pass straight through the atmosphere unimpeded. There is a range of wavelengths, referred to as the **atmospheric window**, in which the atmosphere is also fairly transparent to thermal radiation. At most other thermal wavelengths, however, this is not the case: most radiation emitted by the Earth's surface is absorbed by atmosphere in no time at all. Indeed, it is the small quantities of the trace gases introduced in Chapter 5 that have the largest effects. These absorbers are often referred to as **greenhouse gases**: in absorbing thermal radiation emitted by the Earth, they are both heating the atmosphere and preventing the radiation from leaving the atmosphere, with the net result of global heating. They are so efficient that, of all of the thermal radiation leaving the top of the atmosphere, only a very small fraction (about 7%) was actually emitted by the surface (all of which will be in the range of wavelengths in the atmospheric window). In contrast, 60% of a beam of solar radiation can pass unimpeded through a clear atmosphere. Of the greenhouse gases, carbon

Figure 7.4 Blue sky on a clear day **(A)** and the orange sky seen at sunrise and sunset **(B)** are both caused by light scattering off atmospheric molecules. (Photos: Jon Shonk.)

dioxide (CO_2) is perhaps the best known. Background concentrations of about 200 parts per million (0.02%) exist, although carbon emissions over the last few hundred years are pushing concentrations towards 400 parts per million. Water vapour (H_2O) is also a greenhouse gas, although it receives far less media attention than carbon dioxide. Methane (CH_4) also warms the atmosphere, as do chlorofluorocarbons or CFCs.

The greenhouse effect of CFCs is of far less concern than the impacts of CFCs on global ozone (O_3) distributions. In the troposphere, ozone is also a greenhouse gas. However, the ozone layer in the stratosphere also interacts strongly with solar radiation. More specifically, it absorbs harmful ultraviolet radiation from the Sun, of wavelengths 200 nm to 300 nm, which is damaging to living cells. The absorption process involves a cycle of chemical reactions resulting in decomposition of molecular oxygen (O_2) to individual atoms of oxygen and their combination with other oxygen molecules to form ozone, after which they break down into their constituent parts again. Despite its low concentrations, the ozone is powerful enough to eliminate all of the ultraviolet radiation. The process of **ozone depletion**, caused by the release of man-made chemicals (mainly CFCs) into the atmosphere that irreversibly destroy the ozone molecules is also a concern, and has led to a net loss of ozone in the ozone layer and the formation of the '**ozone hole**' over the Antarctic.

7.4 The Effect of Clouds

To finish our analysis of the Earth's radiation budget, we must also consider the effects of clouds. We briefly alluded to the greenhouse properties of water vapour in the previous section. In its vapour form, water has a strong interaction with thermal radiation, but a small interaction with solar radiation. When it condenses to form either liquid droplets or ice crystals, however, the nature of the interactions changes. Clouds are strong absorbers and emitters of thermal radiation, implying that the presence of clouds results in enhanced emission of radiation towards the surface. In other words, a layer of cloud can provide a sort of greenhouse effect of its own. This can be seen at night; the temperature on a clear night often falls much lower than that on a night with a layer of low cloud present. This is because the cloud is absorbing thermal radiation and emitting some of it back towards the surface.

Clouds also absorb solar radiation energy to an extent, although scattering is again far more important. The bright white forms that clouds take in the sky are due solely to scattering processes. Unlike molecules in the atmosphere that scatter light in all directions, cloud droplets scatter light preferentially forwards. This means that the scattered light appears white, and brighter when viewed near the direction of the Sun. The strong forward scattering gives rise to back-lit clouds appearing to have a very bright region around the edges – the 'silver lining' that every cloud is stated to have (Fig. 7.5).

Via scattering, clouds have a significant effect on solar radiation in the atmosphere by reflecting a fair fraction of the incident radiation back out to space again. The result of this is a cooling of the Earth's surface, as solar radiation that would otherwise heat the surface has been blocked out. This is in stark contrast to the warming of the surface through clouds' emission of thermal radiation. In fact, there is a fine balance between this cooling and warming – indeed, different cloud types can have different effects. Low-level clouds tend to warm the surface, as they emit more thermal radiation to the surface than the solar radiation they block out; high-level clouds tend to cool the surface, as they still block out solar radiation, but have little warming effect on the surface.

Figure 7.5 The 'silver lining' of a cloud is sunlight that has been scattered forward when the Sun is behind it. (Photo: Jon Shonk.)

Water in the atmosphere can also generate a number of fascinating optical effects. When water droplets grow to the size of raindrops, scattering processes become geometric: in other words, the droplets interact with light as if it were within a glass sphere. Sunlight on a raindrop passes in and refracts, then reflects off the side of the drop away from the Sun, before refracting once more upon leaving the drop. The angle through which the light is redirected by this process (**refraction**) has a fixed value of 138°, but with slight variation depending on the wavelength of the light. This means that the different wavelengths will be spread out during the two refraction steps, a process called **dispersion**. The longer wavelengths (red) are refracted through smaller angles than the shorter wavelengths (blue), giving us a rainbow (Figs 7.6A, 7.7A). The rainbow will always appear as a 42° arc around the point directly opposite the direction to the Sun, and is only apparent when there is a large enough concentration of water droplets in the appropriate position in the sky. If the Sun is more than 42° above the horizon, a rainbow will rarely be discernible.

A variety of types of optical phenomena occur when solar radiation interacts with ice crystals, many of which are dependent on the crystal shapes that are present. If an ice cloud

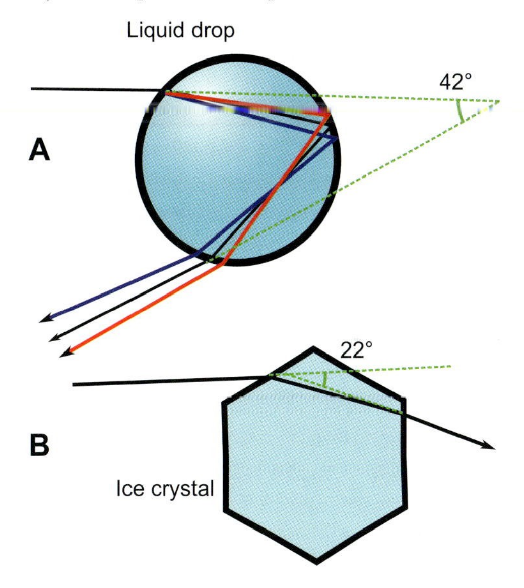

Figure 7.6 The passage of light through (**A**) a raindrop and (**B**) an ice crystal. Light through a raindrop is deflected by 138°, creating a dispersed 42° halo (a rainbow). Through an ice crystal, light is deflected by 22°.

Figure 7.7 The interaction of light with liquid droplets and ice crystals creates a number of striking effects: (**A**) a double rainbow; (**B**) a 22° halo with sundogs to either side of the Sun. (Photos: Michelle Cain (A), NOAA Photo Library/Grant W Goodge (B).)

consists of flat, plate-like ice crystals, a **halo** can sometimes be seen. This halo is located around the Sun (or indeed the Moon) at a fixed angle of 22° – the angle that incident light is refracted through when passing between the opposite flat faces of the ice crystals (Figs 7.6B, 7.7B). If the plate-like crystals are all aligned horizontally, the 22° halo can also be accompanied by **sundogs** (or **parhelia**). These are bright patches of scattered sunlight found either side of the Sun, just outside the 22° halo, and are much more notable when the Sun is near the horizon.

In total, the Earth's radiation budget is far more complicated than just balancing incoming solar radiation with outgoing thermal radiation. Similarly to the hydrological cycle, there are many different streams of radiation connecting all the aspects of the Earth–atmosphere system – the clouds, the surface, the atmosphere itself. The exact global average radiation transferred between these components remains uncertain. Many attempts have been made to determine a set of values for different aspects of the Earth's radiation budget, often using sophisticated computer models in combination with measurements from ground stations or satellites. A typical example of a radiation budget is shown in Figure 7.8. We see that, of the global average value of incident solar radiation, here quoted as $342\,W\,m^{-2}$, approximately $107\,W\,m^{-2}$ is reflected back to space, giving a total albedo of the Earth–atmosphere system of about 0.33. This is over twice that of the Earth's surface alone, showing the radiative impact of the atmosphere and its clouds. This is balanced by a total of $235\,W\,m^{-2}$ of thermal radiation escaping back to space, emitted from the atmosphere, the clouds and, to a lesser extent, the surface via the atmospheric window.

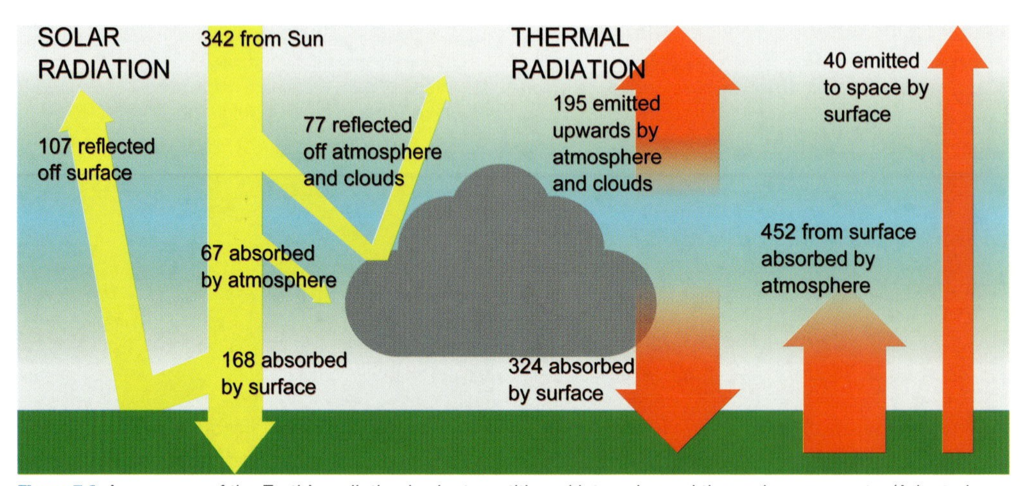

Figure 7.8 A summary of the Earth's radiation budget, partitioned into solar and thermal components. (Adapted from Kiehl and Trenberth, 1997.)

8 Hot and Cold

So the Earth is a system in equilibrium: averaged over a long period of time, it gains as much energy from the Sun as it loses via emission of thermal radiation. If this were not the case, the imbalance would result in a runaway cooling or warming as the Earth–atmosphere system tried to regain equilibrium. But this is far from the end of the radiation story; we know from experience that radiation is not uniformly distributed over the entire surface of the Earth at all times. If it were, the Earth would probably be quite a boring place. Uniform solar heating and terrestrial cooling would generate similar temperatures worldwide and suppress a great deal of the weather systems we experience. Fortunately, the distribution of incoming solar radiation is variable, on scales of both space and time. Local changes in the energy balance result in local warming and cooling as the Earth–atmosphere system constantly struggles to maintain equilibrium. When this solar radiation is absorbed by the Earth, it leads to great contrasts in temperature – and from these contrasts, as we will later discover, all sorts of exciting weather can be created.

8.1 Surface Temperature

When, in conversation, we describe the weather as being either 'hot' or 'cold', we are usually talking about the temperature of the air in the lowest few metres of the atmosphere, often referred to as the **surface temperature**. As we spend most of our time in this part of the atmosphere, surface temperature is an important aspect of our lives; in combination with the presence or absence of cloud and rain, temperature is another great division between fine weather and unpleasant weather. Extreme values of surface temperature, both hot and cold, can be very damaging to our health.

Surface temperature is primarily driven by the emission of thermal radiation from the Earth's surface. A fraction of the incoming solar radiation (defined in terms of the surface albedo) is absorbed by the ground. This warms up the ground and increases the amount of thermal radiation it emits. Much of this emitted energy is rapidly absorbed by the atmospheric gas molecules, implying that most transfer of radiation happens between the surface and only the lowest reaches of the atmosphere.

To change the surface temperature, we must change the balance of the radiation budget at the Earth's surface. If the amount of incoming solar radiation increases, for example, the surface warms up to restore the equilibrium, in turn heating the lower atmosphere. The energy transfer in Figure 7.8 signifies values averaged over long periods of time; if these fluxes were constant, the surface temperature would not change. If the amount of incoming radiation (solar plus any thermal radiation from low-level clouds) exceeds the amount of thermal radiation being emitted by the surface and lower atmosphere, the energy stored in the lower atmosphere increases and

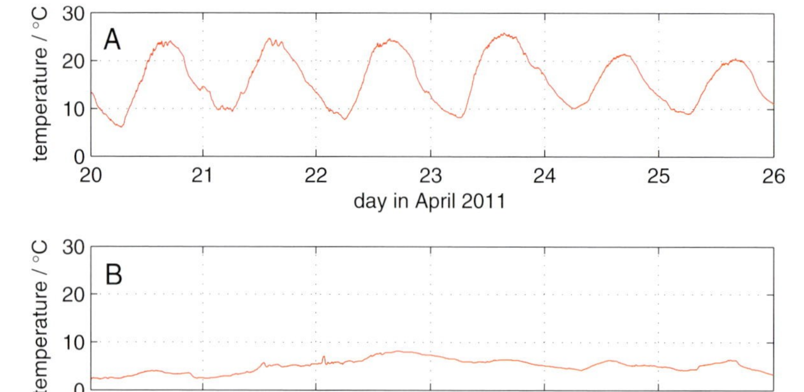

Figure 8.1 Records of temperature at Reading, UK, over (**A**) a period of largely clear days and nights (20 to 25 April 2011) and (**B**) a period of largely cloudy days and nights (26 to 31 December 2010). The dates are marked at midnight on the start of that day. (Data: Department of Meteorology.)

the surface temperature rises. Conversely, if there is more thermal radiation escaping the surface and lower atmosphere than there is coming in, the surface temperature falls.

Over the course of a 24-hour period, strong cycles in incoming and outgoing radiation can create large swings in temperatures between day and night. On a clear, calm day, as the Sun rises and tracks across the sky, the surface is heated and the atmosphere starts to warm up. As the Sun rises higher in the sky, the solar radiation becomes more intense and the heating increases. The incoming solar radiation is greater than the outgoing thermal radiation, even during the afternoon when the surface is at its warmest. The maximum temperature is usually reached during early-to-mid-afternoon, at about 15:00 local time. Once the Sun sets, the source of incoming solar radiation is removed. But thermal energy is still emitted from the surface throughout the night, gradually cooling the lower atmosphere. Under calm conditions, this cooling can continue all night, resulting in a minimum temperature just before dawn. Such strong diurnal cycles in temperature are usually not as pronounced on a cloudy day, as clouds block out solar radiation during the day. However, they tend to keep night-time temperatures higher on account of their emission of thermal radiation. Figure 8.1 shows variations in temperature recorded over a period of sunny days (A) and cloudy days (B).

8.2 Adiabatic Ascent

When the air near the Earth's surface is warmed, however, it does not just sit there at the surface waiting for nightfall so it can cool down again. Radiation is not the only way energy and heat is transported through the Earth–atmosphere system; there is a second method called **convection** that allows warm, moist air near the surface to be transported aloft. Convection is a crucial part of atmospheric circulation and the generation of clouds.

The heating of the Earth's surface by the Sun is far from uniform. Over an area of a few square kilometres, there is likely to be a range of different surfaces, all with different albedos

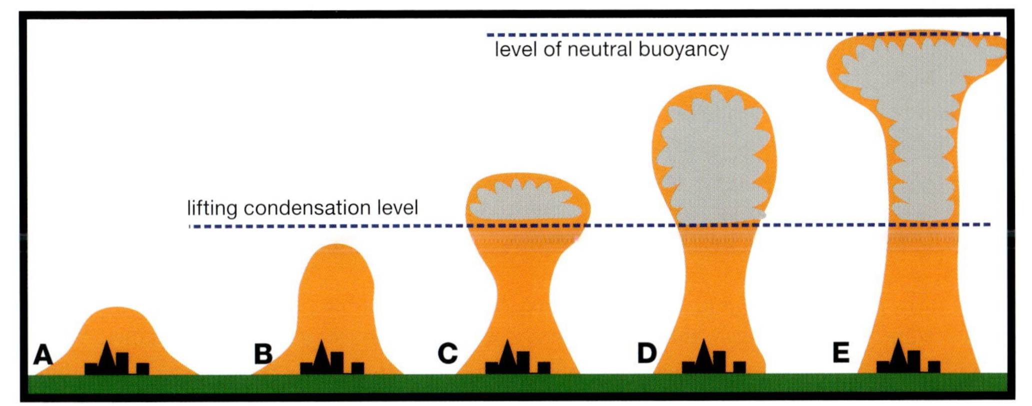

Figure 8.2 An ascending plume of warm air is called a thermal (**A**). Less dense than the surrounding air, the warmer air has a tendency to rise (**B**). As it breaches the lifting condensation level (**C**), cloud begins to form. The plume continues to ascend and form cloud (**D**) until it reaches the level of neutral buoyancy (**E**).

and emissivities. This can result in local differences in surface temperature, and hence variability in air density. Consider a mass of moist air (usually referred to as a **parcel of air**) just above the Earth's surface that has been warmed slightly more than its surroundings. For example, it may have been resting over a ploughed field or an urban area. At a fixed pressure, its higher temperature implies a lower density, hence a tendency to rise, as it is positively buoyant. This rising bubble of warmer air is called a **thermal** (Fig. 8.2).

As the parcel of air rises, the air pressure within it decreases. This means that the parcel must expand and cool. If we assume that no heat enters or leaves the parcel, it will ascend **adiabatically** and cool by 9.8°C for every kilometre it travels upwards. This constant rate of cooling with height is referred to as the **dry adiabatic lapse rate**. If the parcel contains an amount of water vapour, as it cools, its relative humidity must rise (also assuming no transfer of moisture into or out of the parcel). If the parcel is still positively buoyant when it reaches saturation, liquid droplets start to condense and a cloud forms. The layer at which condensation starts to occur is called the **lifting condensation level**. The parcel may still continue to ascend, possibly building up deep, convective clouds. However, as the formation of droplets releases latent heat, the cooling rate of the parcel reduces to about 6°C km^{-1} (the **saturated adiabatic lapse rate**). The value of the saturated adiabatic lapse rate is not constant, but dependent on both moisture concentrations and temperature. All the time that the parcel's surroundings are cooler than the parcel itself, it will continue to rise. However, if this is not the case, the parcel cannot rise any further, nor can its water vapour condense and form clouds any more. The level at which a parcel of air is no longer free to rise is the **level of neutral buoyancy**.

8.3 Clouds, Fog, Dew and Frost

Comparing the observed rate of change of temperature of the atmosphere with height (often referred to as the **environmental lapse rate**) to the dry and saturated adiabatic lapse rates gives us an indication of the **stability** of the atmosphere (Fig. 8.3). In a stable atmosphere, the environmental lapse rate is smaller than the adiabatic lapse rates and hence convection is suppressed. In an unstable atmosphere, the environmental lapse rate is larger than the adiabatic lapse rates. In this instance, convection is free to take place and deep clouds could form that span the entire depth of the troposphere. Often, however, the atmosphere is neither entirely stable nor unstable throughout its depth, but contains a number of stable layers that act as a cap to convection, halting vertical motion in its tracks. A **temperature inversion** is an example of a strong stable layer – a layer where temperature increases with height. The entire stratosphere is therefore a stable layer, with the tropopause marking the base of a temperature inversion. For this reason, tropospheric convection is incapable of penetrating far into the stratosphere.

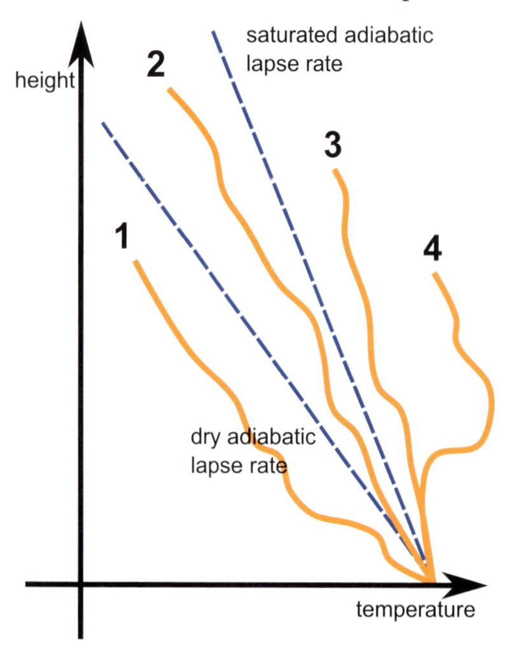

Figure 8.3 Comparing the environmental temperature lapse rate (solid orange lines) with the dry and saturated adiabatic lapse rates (dashed blue lines) gives an indication of the atmospheric stability. Profile 1 is stable, profile 2 is conditionally stable, profile 3 is unstable. Profile 4 is unstable with a temperature inversion.

Stability is therefore an important factor in determining the shapes of clouds. Unstable conditions allow convection to occur rapidly, with the development of tall, deep clouds, such as cumulus and cumulonimbus. Such clouds are referred to as **convective clouds**. If they grow deep enough, convective clouds can bring heavy, showery rain and thunderstorms (see Chapter 12). A stable layer, however, will quickly put a stop to the cloud's vertical growth. This can give rise to vast layers of cloud, referred to as **stratiform cloud**. This could be in the form of a layer of individual cumulus cells (stratocumulus) or a continuous sheet of cloud (stratus). Stratiform clouds also form along weather fronts, where air ascends up a shallow, sloped surface (see Chapter 10).

Water vapour can also condense to form **fog**. On a clear night, a combination of radiative cooling and conduction of heat into the surface can lead to the development of a layer of cold air directly above the surface. If this air is moist and the cooling is strong enough, condensation can occur in the form of a shallow layer of fog. Fog formed by this mechanism is called **radiation fog** (Fig. 8.4). Once a layer of radiation fog has formed, further radiative cooling from the layer top can deepen the cold air layer, thickening the layer of fog. Radiation fog tends to form on cold winter nights and usually disperses at sunrise, when solar radiation heats up the cold air layer and the fog evaporates. A layer of fog can also form when

Figure 8.4 Radiation fog leading to low visibility in Croydon, UK, on 9 May 2000. (Photo: Jon Shonk.)

a mass of moist air is transported over a much colder surface. This also cools the air from below, again resulting in condensation. Such layers of fog are referred to as **advection fog**.

Along with radiation and convection, there is a third method by which energy can be transferred: **conduction**. This only happens to any great extent where the lowest molecules of the atmosphere meet the surface. However, as the temperature of the surface can often fall several degrees colder than the air above it, conduction is a powerful method of removing heat from the atmosphere. Air in direct contact with the surface can be cooled rapidly, resulting in condensation occurring on the surface itself in the form of **dew**. If the **dewpoint temperature** of the air is below freezing, the condensing water vapour freezes onto the surface as ice crystals, forming **frost**. Dew and frost only tend to form when there is little wind and air remains in contact with the ground for a long time before being carried away.

If fog forms at temperatures below 0°C, it can consist of supercooled liquid droplets. Such fog is referred to as **freezing fog** and will quickly freeze onto any exposed cold surface, creating spectacular crusts of ice called **rime** (Fig. 8.5B). If the supercooled droplets become large enough, they can fall as **freezing rain**, which can quickly encase surfaces in much heavier, thicker crusts of ice (Fig. 8.5A). These can cause damage to buildings and trees, and bring down power lines.

Figure 8.5 The results of two ice storms that hit Norman, Oklahoma, in December 2007 (**A**) and January 2010 (**B**). The crust of transparent ice on the tree branches was a result of freezing rain; the spiky rime coating was caused by freezing fog. (Photos: Daniel Peake (A), John Lawson (B).)

8.4 The Spherical Earth

So, daily cycles of incoming solar radiation affect the surface temperature, which can lead to the development of clouds by condensation and change the weather conditions over the course of a 24-hour period. However, we do not expect this diurnal cycle to affect the weather in the same way on all parts of the Earth's surface. For a start, in the tropics, surface temperatures are much higher than in polar latitudes. These spatial variations are, of course, brought about by the fact that the Earth is (roughly) spherical, and are caused by the two effects summarised in Figure 8.6.

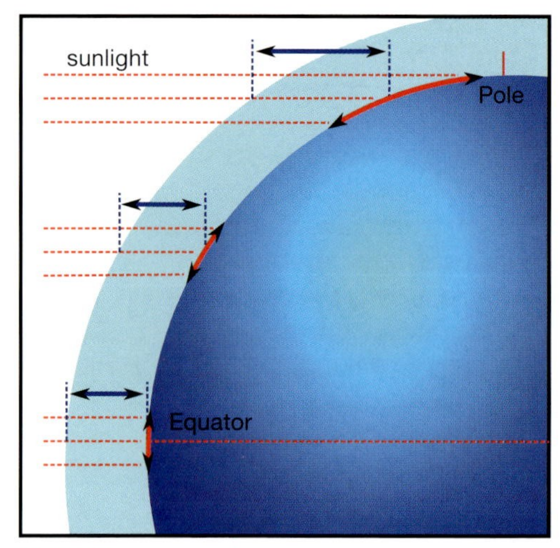

Figure 8.6 In the tropics, the incoming solar radiation is spread out over a much smaller area and must travel through a shorter path of atmosphere than in the polar regions. This results in higher temperatures in the tropics. Note that the atmosphere here is not to scale.

Firstly, the angle at which incoming sunlight hits the ground changes with latitude. When the Sun is directly overhead, the incident solar radiation illuminates the surface from directly above, and therefore a single square metre of radiation illuminates a square metre of the surface. As the Sun falls lower in the sky, the same square metre of radiation will become increasingly spread out over a larger area, as the angle at which the solar radiation hits the surface changes. At an angle of 45° from vertical, the same amount of radiation illuminates 1.41 m^2 of surface; at 60°, it covers 2 m^2; at 85°, it is spread out over 11.5 m^2. During a day in the tropics, the Sun tracks much higher across the sky and can therefore deliver a larger amount of solar radiation to heat the surface. At higher latitudes, it tracks across the sky at much larger angles from the vertical, hence delivering much less radiation to the surface.

A second effect is caused by the depth of the atmosphere. When the Sun is overhead, its rays must travel directly through the depth of the atmosphere to heat the surface. However, when it is low in the sky, its rays must pass through the atmosphere at a much shallower angle. This means that it must pass through more of the atmosphere, allowing more extinction to occur by absorption and scattering. In combination, these processes result in there being far more solar radiation reaching the surface in parts of the Earth where, at local noon, the Sun is highest in the sky – and surface temperatures in the tropics being, on average, tens of degrees warmer than the poles. Figure 8.7 shows the global mean distribution of surface temperature, averaged over the whole year. Average temperatures in the heart of the tropics are between 25°C and 30°C, with temperatures decreasing gradually towards the poles. The Antarctic is, on average, several tens of degrees colder than the Arctic (–50°C as opposed to –20°C).

We must also take into account the swing of the seasons. The polar axis around which

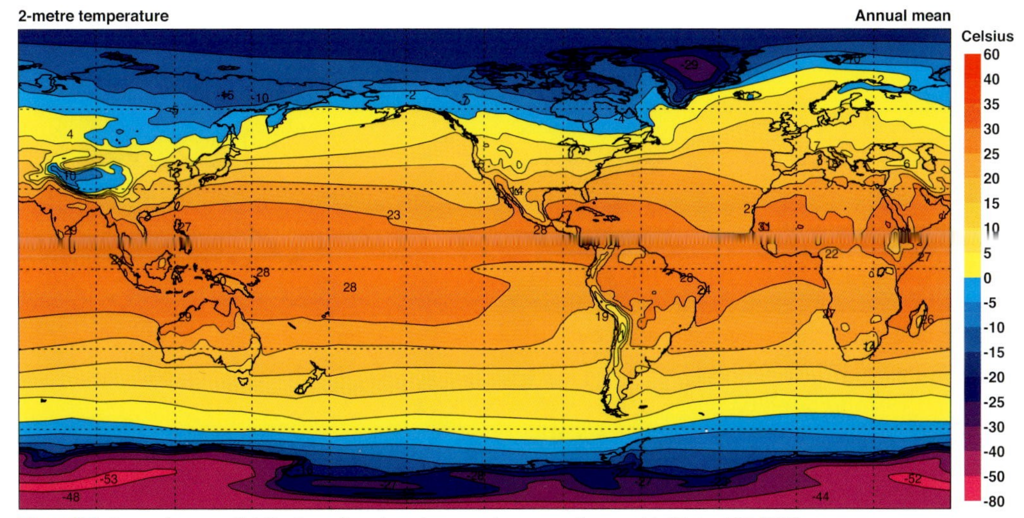

Figure 8.7 Annual mean distribution of surface temperature over the world. (Data: ECMWF.)

the Earth rotates is inclined at about 23.5° off the line perpendicular to its orbital plane. This means that, as it moves around the Sun, each hemisphere of the Earth spends half an orbit tilted towards the Sun, increasing its exposure to solar radiation. In the mid-latitudes, we associate summer with warm, fine weather and winter with cold, stormy weather. In the tropics, where warming varies little over the course of the year, the seasons are marked by other weather tendencies – usually the transition between wet and dry seasons.

As with the diurnal cycle, seasonal cycles in surface temperature are also brought about by temporary imbalances in surface energy budget, but on much longer timescales. During the summer, the Sun spends more time higher in the sky, and day length increases. These effects increase the amount of solar radiation that reaches the surface. The surface therefore heats up, which in turn heats up the atmosphere as summer continues. As the days become shorter again, and the Sun becomes lower in the sky, the solar radiation reaching the surface decreases again, and the surface cools by a similar mechanism.

Although we experience strong diurnal cycles in temperature in the lower reaches of the atmosphere, the changes in temperature from day to night in the upper troposphere are small. Seasonal cycles in temperature, however, affect the whole depth of the troposphere. As the volume of the troposphere is so vast, it takes time to respond to these cycles, resulting in a lag between day length and surface temperature. In the northern hemisphere, the longest day is 21 June, but the warmest summer months are usually July and August; the shortest day is 22 December, but the coldest winter months are usually January and February. The ocean is even more resistant to temperature change, with an even slower response to seasonal cycles in temperature – warmest and coldest sea surface temperatures in the northern hemisphere typically occur three months after the longest and shortest day.

8.5 Variation of Total Solar Irradiance

Diurnal and seasonal cycles in the weather are something we are very familiar with. However, the Earth also experiences 'seasons' that stretch over much longer periods of time, driven by changes in the amount of solar radiation it receives. While not relevant to year-to-year weather conditions, these long-term warming or cooling events can have an impact on the Earth's climate. For a start, the Sun does not provide us with a constant supply of energy over the course of its 11-year solar cycle. In its active phase, marked by an increased number of sunspots, the amount of radiation reaching the Earth can increase by about 5 W m^{-2} with respect to its inactive phase. This is only a small change from the 1,360 W m^{-2} the Sun usually provides, but links have been suggested between this solar cycle and global weather patterns. The strength of the solar cycles also varies over much longer periods, sometimes resulting in a chain of solar cycles where sunspot count remains low and there is little increase in total solar irradiance. From 1650 to 1700, the Sun remained virtually inactive during a period now known as the **Maunder Minimum**. This 50-year drop in solar activity coincided with a drop in global temperatures (the so-called **Little Ice Age**), although scientists now think that this cooling was due to volcanic activity rather than the reduced total solar irradiance.

Changing the distance between the Earth and the Sun also has the same effect of changing the total solar irradiance. The Earth does not orbit the Sun in a perfect circle, but has an elliptical orbit (Fig. 8.8). When the Earth is furthest from the Sun, or at **aphelion**, the incoming solar radiation is therefore at its lowest; when it is nearest, or at **perihelion**, the incoming radiation is at its highest. Perihelion and aphelion occur regularly, six months apart, each and every year. Perihelion occurs during January; aphelion occurs during July. The swing in total solar irradiance caused by the eccentricity of the Earth's orbit is much greater than that from the solar cycle, with a magnitude of about 40 W m^{-2}.

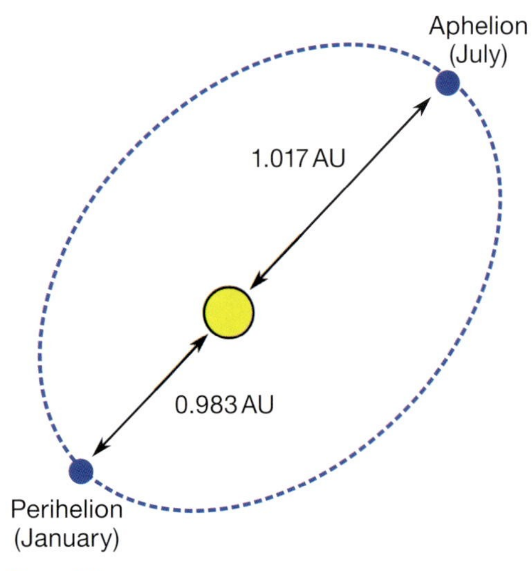

Figure 8.8 The Earth's orbit is elliptical. This varies the amount of solar radiation it receives throughout the year. The perihelion and aphelion distances are in astronomical units, where 1 AU is equivalent to 1.496×10^8 km, the mean radius of the Earth's orbit.

8.6 The Earth in Non-Equilibrium

The variations in surface temperature that are such important factors in our day-to-day weather are actually a part of the Earth–atmosphere system's constant battle to balance its radiation budget and reach equilibrium. While the variability of the incoming radiation budget over short periods of time is constantly changing, the long-term goal of the Earth–atmosphere system is to achieve equilibrium in spite of the many perturbations that are applied to it, both natural (for example, the solar cycle) and man-made (such as global warming – see Chapter 16).

In short, if we average the Earth's radiation budget over a long period of time, we find that, globally, the incoming solar radiation balances the outgoing thermal radiation. However, on a more regional scale, we find that this is not the case. In the tropics, the amount of incoming solar radiation (averaged over many years) is greater than the amount of outgoing thermal radiation. Conversely, the incoming solar radiation is lower than the outgoing thermal radiation at the poles (Fig. 8.9). In other words, there is an excess of energy in the system in the tropics, and a deficit of energy at the poles. For this to be true, the tropics should be becoming steadily warmer and warmer with time, while the poles should be becoming cooler and cooler. But temperature records have been kept for centuries and there is certainly no indication that such regional trends in temperature are occurring. So there must be a way that this imbalance of energy is cancelled out.

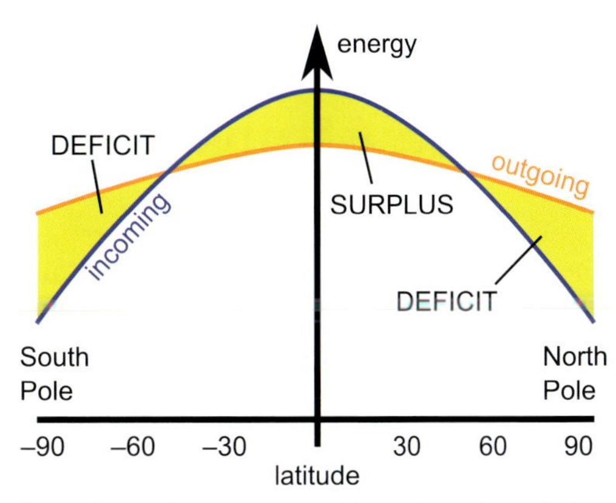

Figure 8.9 Latitude averages of incoming solar radiation (blue) and outgoing thermal radiation (orange). There is an energy excess in the tropics and an energy deficit at high latitudes.

9 The Atmosphere in Motion

The air contained in the atmosphere is constantly in motion. Energy provided by the Sun powers this motion by heating the air and driving circulations. We experience the horizontal flow of air as wind, and observe the presence of vertical flow by the formation of clouds. Air can travel hundreds of kilometres in a day and complete a journey through the height of the troposphere in about a week, transporting and redistributing both energy and moisture.

In the troposphere, we have already seen how local gradients in pressure, temperature and density can drive circulations of air and form cloud. However, these local-scale circulations that span a kilometre or two are far from the end of the story. Air circulates in the atmosphere on a wide range of scales, from the smallest eddies and vortices a few centimetres across up to vast circulations that can span thousands of kilometres. It is via global circulations that the energy excess of the tropics is redistributed, filling in the energy deficit found over the poles.

9.1 Highs, Lows and Circulation of Air

Of the many meteorological variables that we watch every day, none has quite as much impact on our day-to-day weather experiences as atmospheric pressure. While weather conditions may sometimes change from one day to the next, it is not unusual for conditions to remain similar over a number of days, giving a spell of either fair or poor weather. These spells are brought about by the passage of areas of high and low pressure, collectively referred to as weather systems. Between these weather systems, horizontal pressure gradients form, and it is these gradients that set the air in motion.

We have already seen that air pressure is related to the weight of the atmosphere above a given point. So, above a region experiencing high pressure, there must be a greater number of gas molecules in the column of air than there are in the column above a neighbouring region experiencing low pressure. If the two columns are at the same temperature, then the density in the high-pressure column must be greater; if the two columns have the same density, then the temperature (and also the volume) of the high-pressure column must be greater. Both instances result in a tendency for gas molecules to travel from the high-pressure column towards the low-pressure column to cancel the pressure difference (Fig. 9.1). The force on a mass of air that pushes it from a high-pressure area to a low-pressure area is the **pressure gradient force**, and is dependent on the strength of the pressure gradient. This force does not just act horizontally; a vertical pressure gradient force, driven by the decrease of air pressure with height, is constantly pushing the air in the atmosphere upwards. Fortunately, this force is cancelled out by the weight of the atmosphere above. When the vertical pressure gradient force and the weight are in equilibrium, the atmosphere is said to be in **hydrostatic balance**.

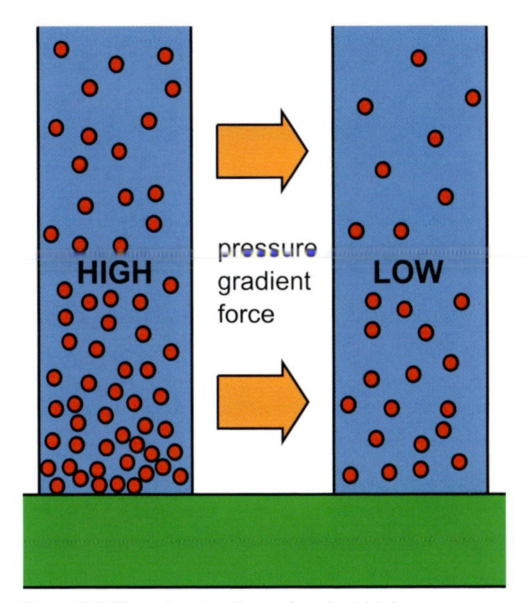

Figure 9.1 There is a tendency for air at higher pressure to move towards an adjacent region at lower pressure. The difference in pressure applies a force to the air, known as the pressure gradient force.

A stream of air from a high-pressure area to a low-pressure area is only a small part of the story. If the wind were restricted to simply blowing horizontally, it would transfer the necessary molecules from the high-pressure area to the low-pressure area, then cease. However, we know that pressure systems tend to last for several days. The reason for their persistence is that the surface winds are actually a small part of a circulating system. When the winds meet in the centre of a low-pressure system, the air they transport cannot just disappear – it has to go somewhere. So the air must ascend. Similarly, in the centre of the high-pressure area, air must constantly be descending to replace the air transported away. In short, the flow of air is continuous: while circulating, it cannot be created or destroyed. In comparison to the magnitude of the horizontal wind speed, vertical wind speeds are very small; typical values are up to 10 cm s^{-1} in contrast to horizontal winds of possibly tens of metres per second high in the troposphere.

This basic circulation occurs in many places in the atmosphere on many different scales. To complete the circulation, we need some sort of return flow aloft to transport the air rising in the low-pressure area back into the descending column in the high-pressure area. However, upper-level flows of air tend to consist of very fast-moving winds which may not necessarily be in the direction required to carry the air where it is needed. Here, regions of convergent and divergent flow can perform the same task.

If the flow is convergent at the top of the troposphere, there is a tendency for flow to be towards the point; in other words, in a horizontal plane, more air is arriving at a point than is leaving it. By **continuity**, this means that air is pushed to descend towards the surface. Similarly, in divergent flow aloft, less air is arriving at the point than is leaving it, which must lead to air being pulled upwards from below. Hence, in the low-pressure area, upper-level **divergence** near the tropopause pulls the air upward from below, causing **convergence** at the surface. In other words, a low-pressure system is characterised by convergence at the surface and divergence aloft. For a high-pressure system, the converse is true, with divergence at the surface and convergence aloft (Fig. 9.2). Indeed, such areas of upper-level convergence and divergence play an important role in the creation and destruction of **mid-latitude depressions**, as we will see later, in Chapter 10.

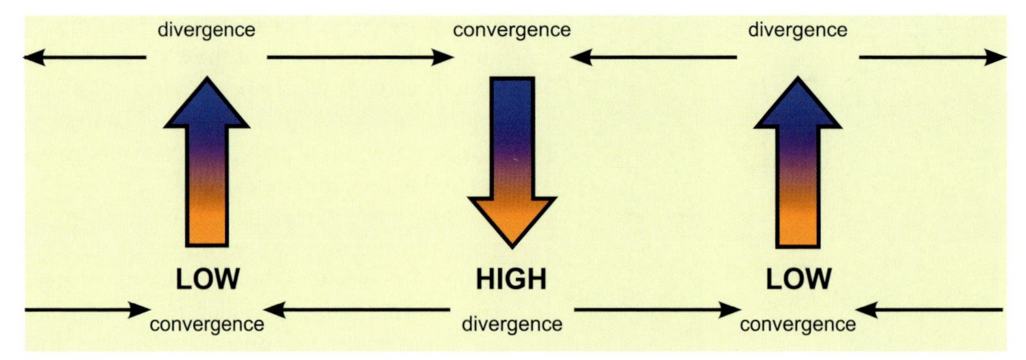

Figure 9.2 An area of high pressure is associated with divergent flow at the surface and convergent flow aloft; an area of low pressure is associated with convergent flow at the surface and divergent flow aloft.

9.2 The Coriolis Effect

There is another force, in addition to the pressure gradient force, that is applied to air in motion. So far, we have not accounted for the rotation of the Earth. If the Earth did not rotate, the flow between high and low pressure would be direct. However, the patterns of winds we see on weather charts show them spiralling out of the high-pressure system and into the low-pressure system. It is the rotation of the Earth that causes the winds to spiral in this way.

Consider an aircraft flying from Oslo to Rome (marked as O and R respectively in Figure 9.3), a journey that is practically due south. The pilot takes off from Oslo and points his aircraft south. But the flight takes about four hours to complete and, as he arrives at the latitude of Rome four hours later, the Earth has rotated an entire sixth of a turn. Although the pilot has flown due south in the correct direction, he ends up missing Rome by a considerable margin, ending up somewhere over the Atlantic Ocean. From the perspective of the pilot, he has flown in a straight line, but from the viewpoint of an observer moving with the Earth, his path will appear to have curved to the right.

Indeed, if he flies along any path in a given direction in the northern hemisphere, his path will always curve to the right. If he crosses into the southern hemisphere, his path will always curve to the left. This deflection caused by the Earth's rotation is attributed to the imaginary **Coriolis force**, named after French engineer Gaspard-Gustave Coriolis. For any moving object, be it an aircraft or a mass of air, the Coriolis force acts at right angles to its motion and is a function of both latitude and the speed of motion. In the northern hemisphere, it deflects the movement of air to the right; in the southern hemisphere, it deflects the movement of air to the left. This causes air to always flow clockwise around an area of high pressure and anticlockwise around an area of low pressure in the northern hemisphere and vice versa in the southern hemisphere. In other words, if an observer backs into the wind in the northern hemisphere, the low pressure is always to the observer's left. This rule is **Buys Ballot's law**.

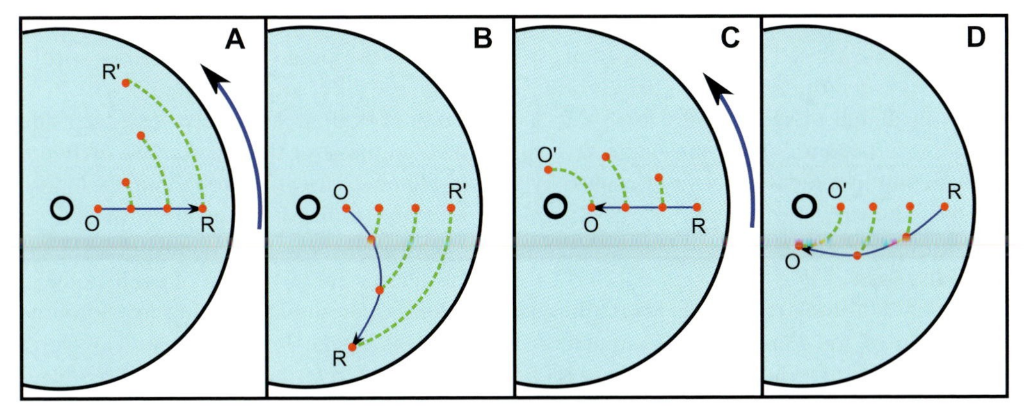

Figure 9.3 The effect of the Coriolis force on an aircraft travelling due south from Oslo to Rome (**A**, **B**) and due north from Rome to Oslo (**C**, **D**). With respect to the pilot, his journey is a straight line (**A**, **C**), but with respect to a viewer on the Earth's surface, his path has curved to the right (**B**, **D**).

So, air travelling from a high-pressure system to a low-pressure system is affected by both the pressure gradient force and the Coriolis force. In the northern hemisphere, the pressure gradient force is always down the pressure gradient; the Coriolis force is always 90° to the right of the direction of motion. If the pressure gradient force and the Coriolis force on an air mass are equal in magnitude and opposite in direction, its motion is said to be **geostrophic** and the mass moves in a direction parallel to the **isobars.** Aloft, winds are often near-geostrophic in nature. In the lower atmosphere, frictional forces between the air and the surface affect this balance, slowing down the winds and causing them to cross the isobars at an angle (Fig. 9.4). The component of wind blowing across the isobars is referred to as the **ageostrophic** wind.

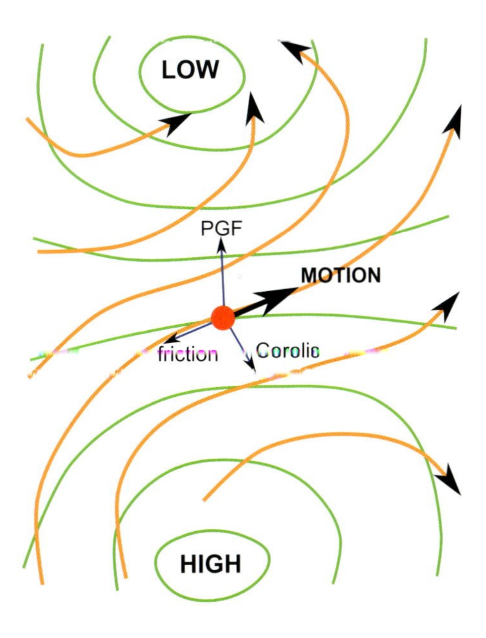

Figure 9.4 Winds (shown in orange) spiral outward from a high-pressure system and into a low-pressure system in the northern hemisphere. The spiralling motion is due to the combined effects of the pressure gradient force (marked as PGF), the Coriolis force and frictional forces between the air and the ground.

It should also be noted that the Coriolis force only affects objects that travel significant distances across the Earth's surface. For example, while the Coriolis force has a significant effect on the journeys of air masses crossing the Atlantic, it has a very small effect on water crossing a bath. The direction of rotation of water draining from a bath is often falsely attributed to the Coriolis force – it is much more dependent on the geometry of the bath and motions in the water caused by removing the plug.

9.3 Hadley Cells

We now have the tools required to give a simple description of the Earth's global circulation and its methods of redistributing the excess of energy deposited in the tropics (see Chapter 8). The tropics receive radiation strongly from the Sun all year, with little seasonal variation. At equinox, the strongest heating is over the Equator. We have seen how a region that is heated more strongly than its surroundings warms the air directly above it, reducing the density of this air and causing it to rise. The same effect happens in the tropics – although over a much larger scale.

The effect is very similar to that of a heater placed in the middle of a room. It heats the surrounding air, creating a rising plume of warm air. When the plume reaches the ceiling, it spreads out horizontally and gradually cools, falling back down towards the floor to be transported back into the column of air being heated. We may therefore expect the entire troposphere to be a single convective circulation, with the heating at the tropics generating vast regions of ascending air that travel polewards and descend at higher latitudes. But this is not the case: the air that has risen to the top of the troposphere over the Equator descends back to the surface at low latitudes (about 30° either side of the Equator). The width of these vast convective circulations – referred to as **Hadley cells** – is used in meteorological circles to define the tropics, as the behaviour of weather within them is very different to elsewhere.

A pair of Hadley cells exists – one each side of the Equator. Over the Equator, we therefore get a region of ascent throughout the entire depth of the troposphere, fed by warm air from the sub-tropics in both hemispheres. This ascent allows the formation of deep, convective clouds, potentially bringing vast amounts of rain. This band of low pressure and low-level convergence – referred to as the **Intertropical Convergence Zone** – appears as a broken band of cloud and precipitation running all around the world. As the seasons change, this band of convergence tracks northward and southward, giving rise to the wet and dry seasons typical to the tropics (see Fig. 9.5).

Away from the Equator, we find subsidence. Despite being on the cooler, descending branch of the Hadley circulation, the subsiding air is still very warm. This is because, as it descends, the increasing air pressure it is subjected to warms it by the dry adiabatic lapse rate. The air is also very dry, as most of its moisture content has rained out during its ascent. As subsiding air also suppresses cloud formation, this results in a vast swathe of the Earth where cloud formation and precipitation are rare. These latitudes contain most of the Earth's deserts – including the Sahara Desert, and also the deserts of Australia. There is also little wind at these latitudes, as the air diverging at the surface leads to vast areas of high pressure, called **sub-tropical highs**. Surface winds in the tropics blow from the sub-tropical highs towards the Intertropical Convergence Zone. These Equatorward winds are referred to as **trade winds**, as their consistency was important to international trade in

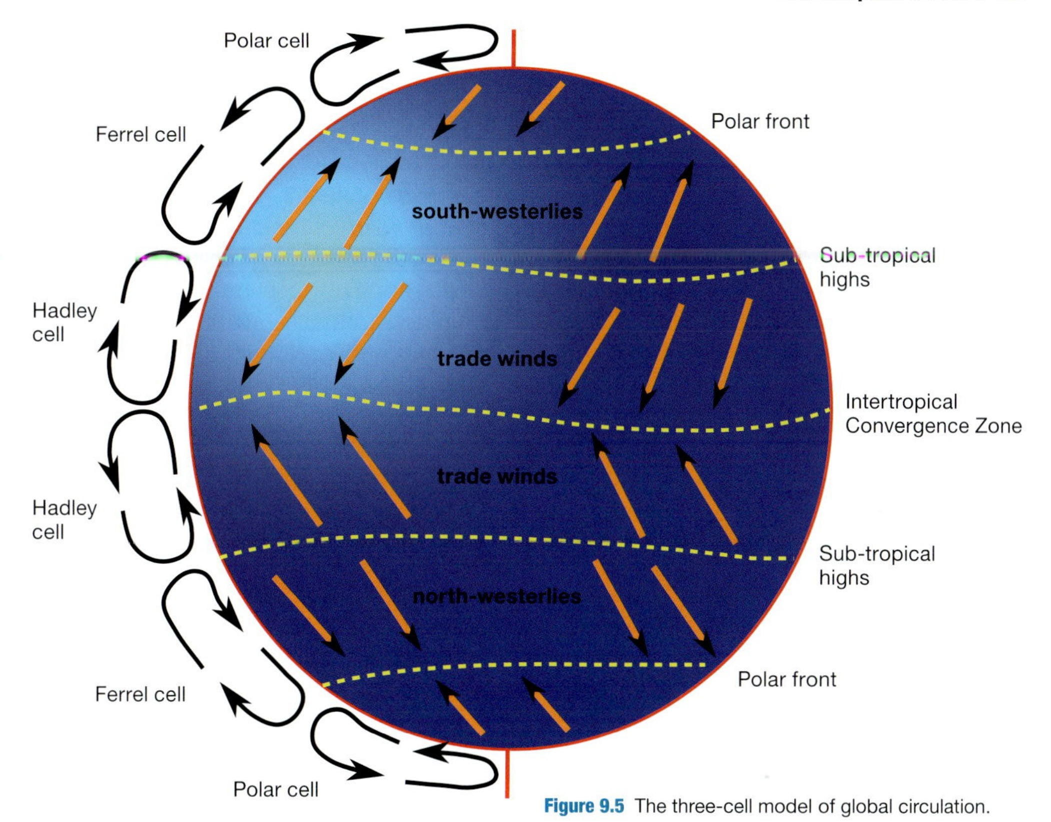

Figure 9.5 The three-cell model of global circulation.

the time of sailing ships. The northerly trade winds in the northern hemisphere and the southerly ones in the southern hemisphere receive an easterly component via the Coriolis effect.

9.4 Heat Transport in the Mid-Latitudes

So in the tropics we have circulation that transports excess heat polewards – although not all the way to the poles. For the deficit of energy over the poles to be offset, this circulation alone cannot be the whole picture. Indeed, the process of the Hadley cell seems to merely move heat around within the region where there is an excess of energy. We need to move the heat contained in this circulation away from the tropics.

It turns out that the Hadley cell is one of three vast rotating cells in each hemisphere. Over the polar regions, the air circulates in a similar way to the Hadley cell. There is an ascending branch at the southernmost limit (about 60° north or south) and a descending branch at the northernmost limit (over the pole). Referred to as a **Polar cell**, it is a much weaker circulation than the Hadley cell, on account of the much smaller amount of energy in the system at such high latitudes. The descending branch of the polar cell leads to high pressure and suppresses cloud and precipitation over the pole.

This leaves a gap spanning the so-called **mid-latitudes** (between about 30° and 60°) that has so far been unaccounted for. This gap has descent on its southern side and ascent on its northern side indicating that, by continuity, surface winds in this region must blow from south to north (deflected towards the east by the Coriolis effect). This cell is often referred to as the **Ferrel cell**, and again its circulations are much weaker than those of the Hadley cell. The weak convergence of the warmer air from the south and the colder air from the north create a boundary of low pressure known as the **polar front**. Perturbations along this polar front give rise to mid-latitude weather systems (see Chapter 10).

However, the Ferrel cell actually does little to explain how heat is transported between the Equator and the poles. Indeed, the presence of the Ferrel cell at all is usually only discernible in long-term averages of winds – at any one time, the flow patterns of the mid-latitudes are typically complex and mask this weak circulation. To describe the heat transfer, we must consider the properties of the air on either side of the mid-latitudes. In the tropics, heat is supplied fairly uniformly over the whole year, so the horizontal gradients in temperature are small. Towards the poles, there is a dearth of incoming energy, so the horizontal gradients in temperature are also quite small here, although obviously with much colder temperatures. Between the two exists a much stronger horizontal temperature gradient, flanked by vast columns of air that are either warm or cold throughout their entire depth.

We know that air pressure decreases roughly exponentially with height up into the atmosphere. This means that, at some point above us, the pressure must be, for example, 500 hPa. The height of this pressure level above the surface is related to both the temperature and density of the column of air below, so will not necessarily be the same everywhere. In a colder column of air, the molecules are much less energetic than in a warmer column. This means that the air in the cold column is denser than the air in the warm column.

This has an effect on the pressure levels. At a given point in the atmosphere, the air pressure is related to the number of molecules above that point. So, in the less dense, warm column, the pressure levels are more spaced out, with the 500 hPa surface being higher up than in the colder column. In polar air, the 500 hPa surface is usually between 4.8 km and 5.0 km up; in tropical air, it is at about 5.8 km to 6.0 km. Therefore, at a given height, the pressure in the warm column is greater than in the cold column, resulting in a pressure gradient force pushing the warmer air poleward. The difference in pressure between the warm and cold columns increases with height, implying that the pressure gradient force is stronger in the upper troposphere. This all results in a southerly flow across the northern hemisphere mid-latitudes and a northerly flow across southern hemisphere mid-latitudes. The Coriolis force deflects these winds towards the east, generating a band of westerly flow that increases in strength with height in both hemispheres. The strongest winds are in the **jet stream**, right up near the tropopause (Fig. 9.6).

It is the combination of the upper-level flow and the polar front that enables the formation of mid-latitude weather systems. The interaction of warmer air from the tropics with the cooler air from the poles during the lifetimes of these weather systems transports energy towards the poles. They can carry warm air masses to very high latitudes, sometimes delivering energy into the polar cells. A more in-depth discussion of the formation of mid-latitude depressions is given in Chapter 10.

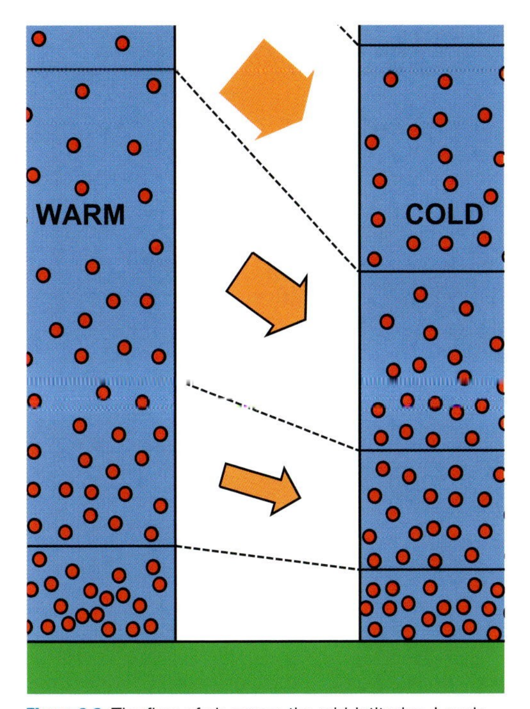

Figure 9.6 The flow of air across the mid-latitudes. Levels of constant pressure slope steeply downward across the mid-latitudes. Air moves along these surfaces of constant pressure towards the poles. As the pressure surfaces slope most steeply near the tropopause, winds here are fastest.

9.5 The Global Circulation

In reality, the passage of weather systems and local circulations masks the simple three-cell circulation patterns. Even so, when averaged over long periods, the patterns can still be identified in climatologies of global circulation. Figure 9.7A shows a map of mean-sea-level pressure for the entire year, averaged over many years. We see a band of low pressure near the Equator marking the Intertropical Convergence. This coincides with the rising edge of the Hadley cells. At their descending branches, we see the sub-tropical highs. These are very persistent features, and many are named for the areas over which they sit: for example, the high pressure over the sub-tropical Atlantic is often referred to as the **Azores High**. The mid-latitudes contain areas of low pressure where the warm tropical air meets the cold polar air, although the regions where these meet tend to be quite smeared out. In fact, the deepest areas of mean low pressure seem to be in swathes across the northern Atlantic and Pacific Oceans. These are referred to as the **storm tracks**, and indicate the typical paths taken by mid-latitude depressions as they track over the sea. Much longer storm tracks are also found in the Southern Ocean, where systems are free to track unimpeded around the ocean without making landfall. Global winds are also shown in Figure 9.7B. We see that winds are strongest where pressure gradients are steepest. The trade winds and the mid-latitude westerlies are most apparent over the oceans, where their passage is unimpeded.

So energy is transported from the tropics, where there is a net excess, to the poles, where there is a net deficit. This transport keeps the Earth–atmosphere system in equilibrium, despite its significant local imbalances.

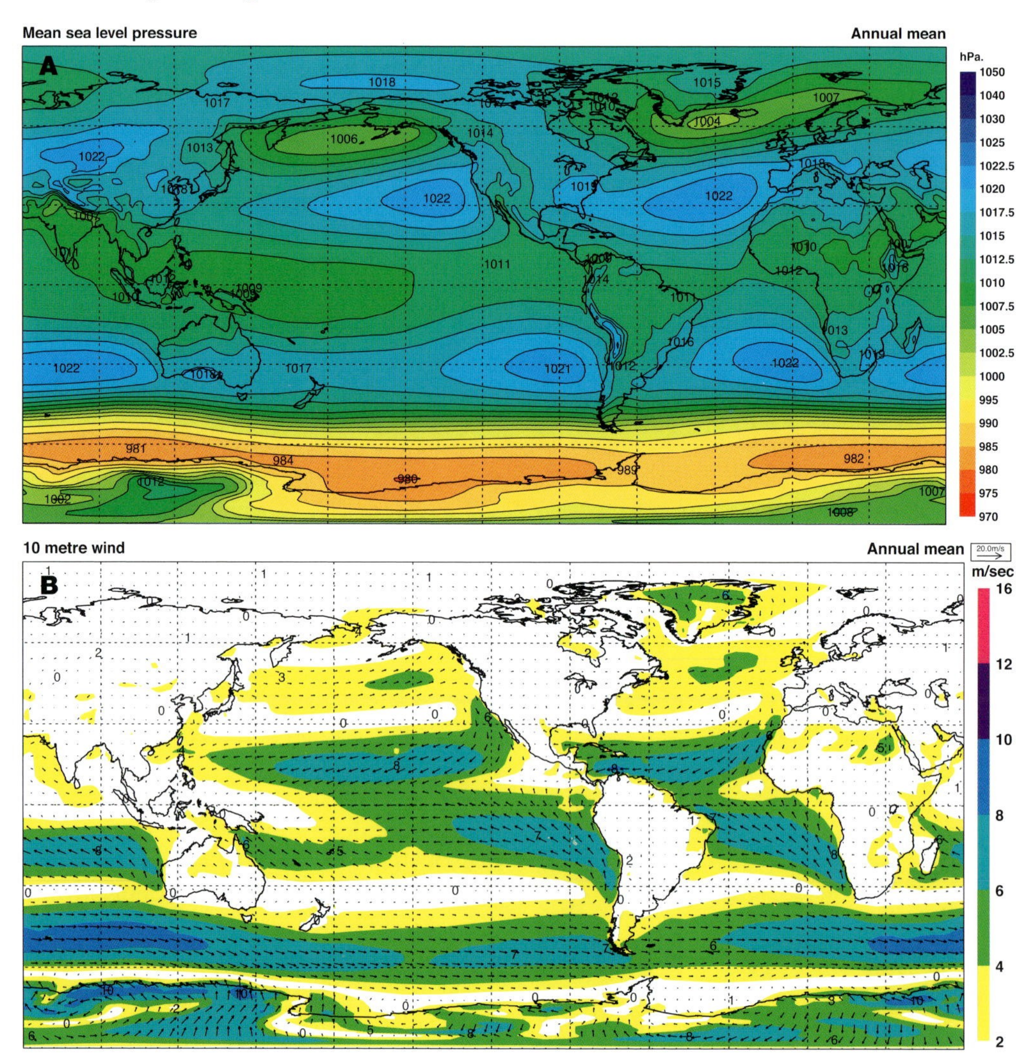

Figure 9.7 Annual mean map of mean-sea-level pressure (**A**) and surface wind speed and direction (**B**). The subtropical highs and storm tracks are clearly evident. (Data: ECMWF.)

During its journey, this energy can be used to generate all sorts of weather. However, the ways it is transported are very different when crossing the tropics and the mid-latitudes and, on account of this, the weather systems that can be produced in these two locations are very different.

10 Mid-Latitude Weather Systems

Weather in the **mid-latitudes** is marked by the procession of regions of high and low pressure around the Earth. High-pressure regions, which bring fair, settled weather, are in a constant battle with low-pressure regions, which bring unsettled weather, clouds and rain. Anyone who has seen a TV weather forecast will have seen this battle being played out.

Mid-latitude weather systems are a small part of the vast global circulation system that ultimately acts to offset the imbalance of energy between the tropics and the poles. They form where tropical air meets polar air and, during their life cycles, mix the air up. This mixing helps to transport some of the excess energy in the tropics towards the poles. Along the way, of course, some of it is used to form swathes of cloud and rain. Mid-latitude weather systems are structured, with most of the cloud and rain they bring centred on bands called fronts – indeed, very different in structure to tropical weather systems, as we will later discover.

10.1 The Westerly Flow

Flanked on one side by the warm tropics and on the other by the cold poles, the intermediate band commonly referred to as the mid-latitudes is marked by the presence of strong north–south temperature gradients. As we saw in the previous chapter, these gradients cause air from the deeper, warm tropical column to flow northwards to the shallower, cold polar column. These poleward winds are deflected to the east in both hemispheres by the Coriolis effect, resulting in a general westerly flow in the mid-latitudes.

These westerly winds are strongest near the tropopause, where pressure surfaces slope downwards most sharply. Up here, winds scream around the Earth at high speed. The fastest winds are found in the **jet stream** – a thin ribbon of very fast-moving air, typically about 100 km wide, but only 2 km deep – winds of over 100 m s^{-1} (220 mph) have been recorded here. The jet stream snakes its way around the mid-latitudes. The strongest winds of the jet stream lie above the polar front (Fig. 10.1). This is the region where the north–south temperature gradient is at its sharpest in the

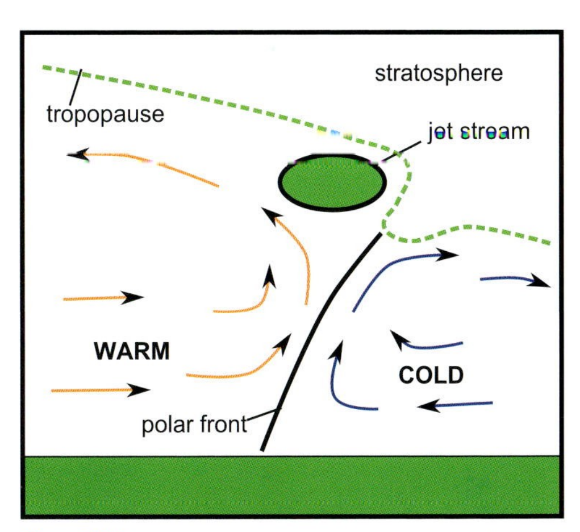

Figure 10.1 Cross-section through the polar front, where warm, tropical air meets cold, polar air. The tropopause is markedly lower in the cold air, and the jet stream runs in the upper reaches of the warm air.

mid-latitudes. In the three-cell model, it marks the convergence line where the poleward branch of the Ferrel cell meets the Equatorward branch of the polar cell. However, as the three-cell circulation is rarely discernible outside the Hadley circulation, it is perhaps more useful to think of the polar front as being the boundary where warm air from the sub-tropics meets cold air from high latitudes. The polar front is certainly not a stationary feature: as masses of warm and cold air impinge upon it, it bends and twists and changes position. Together, the temperature contrasts of the polar front and variations in the speed and direction of the winds in the jet stream lead to the generation of mid-latitude weather systems.

A typical map of mid-latitude weather contains areas of high pressure called **anticyclones** and regions of low pressure called **depressions** (Fig. 10.2). There are no particular pressure thresholds to define an anticyclone or a depression; the only requirement is a higher or lower pressure than the surroundings. They usually appear on pressure maps surrounded by at least an isobar or two, although well-established anticyclones or severe depressions can be several tens of hectopascals higher than the surrounding area. Long extensions of high pressure are called **ridges**; long extensions of low pressure are called **troughs**.

10.2 Anticyclones and Air Masses

The flow of winds between anticyclones and depressions in the mid-latitudes follows exactly the same principle as in all other parts of the global circulation: depressions (low pressure) are marked by converging air at the surface and winds spiralling inwards in an anticlockwise direction in the northern hemisphere (clockwise in the southern hemisphere). By continuity, this means that air in the heart of a depression is ascending, with a divergent flow aloft. Conversely, anticyclones (high pressure) are marked by diverging air at the surface and winds that spiral outwards in a clockwise direction in the northern hemisphere (anticlockwise in the southern hemisphere), with descending air at their centres and convergence aloft.

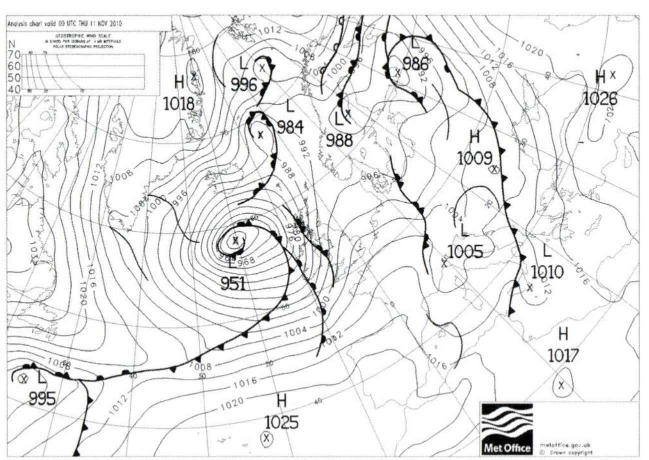

Figure 10.2 Pressure map of Europe and the North Atlantic on 11 November 2010 at 00:00 UTC. (Image: Met Office.)

The descending air at the core of an anticyclone suppresses cloud formation, resulting in an area of largely settled weather. The pressure gradients in the centre of an anticyclone tend to be weak, so winds are often light. In the summer, anticyclonic conditions can bring long spells of hot, dry weather, particularly if they contain sub-tropical air. In the winter, anticyclonic conditions can be much colder, with clear skies and calm conditions resulting in very cold nights.

An anticyclone can be classified as either a cold-core anticyclone or a thermal or warm-core anticyclone. A warm-core anticyclone contains a deep column of descending air which can extend through the entire troposphere. As air in this column descends, increasing atmospheric pressure warms it dry adiabatically. Despite their title, warm-core anticyclones can still bring very cold weather in winter. Indeed, most anticyclones that affect coastal areas of the mid-latitudes (such as western Europe, the US coasts and eastern Asia) are warm-core in nature, as are the sub-tropical highs. Descent in a warm-core anticyclone can often create a strong temperature inversion a kilometre or two above the ground. If warm, moist air becomes trapped underneath this inversion, widespread stratiform cloud may develop and persist for several days in the calm winds in the centre of the system. Such conditions frequently occur in the sub-tropical highs over the oceans, where layers of marine stratocumulus can form and persist for months on end. Dull weather associated with anticyclonic conditions is often referred to as **anticyclonic gloom**.

In contrast, **cold-core anticyclones** are much shallower features, with descending air only in the lowest few kilometres of the troposphere. They usually form in winter over high-latitude snow-covered land surfaces, where a great deal of the incoming solar radiation is reflected, and hence thermal cooling over a wide area results in rapid cooling of the air above, causing it to descend. Noted examples of such cold-core continental anticyclones often form in winter over Siberia and northern Canada.

A well-developed, deep anticyclone can be a very persistent feature, possibly sticking around over an area for several days or even weeks. Their longevity can result in air being trapped inside them for a long time. During this time, mixing of the air can smooth out local differences in humidity and temperature, resulting in a vast amount of air with very uniform properties, referred to as an **air mass**. The properties of air masses depend on the region in which they form: in the mid-latitudes, there are five main types of air mass that influence the weather, and each typically brings different weather conditions.

Typical origins of air masses in the northern hemisphere are shown in Figure 10.3. Warm air masses generated in the sub-tropical highs are referred to as tropical air masses. If they formed over the land, they are classed as **tropical continental**; if they formed over the ocean, **tropical maritime**. When transported into the mid-latitudes, tropical air masses bring warm weather, with the maritime air mass bringing much more moisture than the continental air mass, leading to more cloud and rain. Air masses that form over higher latitudes are referred to as polar air masses, again subdividing into polar continental and polar maritime, depending on their source. A **polar continental air mass** is dry and cold, and can bring very cold weather in winter.

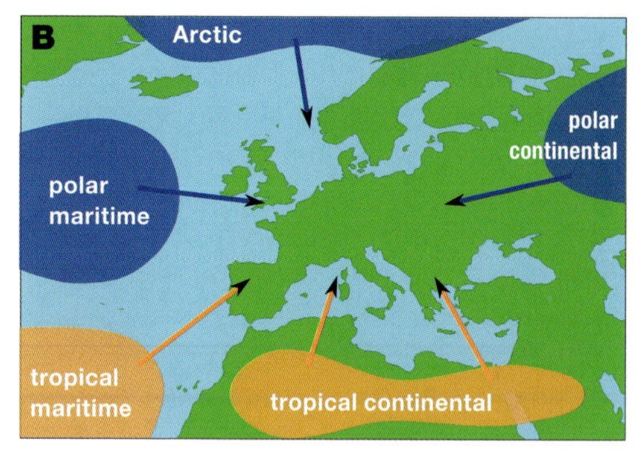

Figure 10.3 The origins of air masses that affect the weather in (**A**) North America and (**B**) Europe.

Polar **maritime air masses** tends to be moister and milder, as its passage across the sea can result in it being warmed substantially more than the continental air does passing over the cold land. A final type of air mass comes from even higher latitudes. These **Arctic air masses**, which form over the polar regions, can bring bitterly cold weather from the north, and also plenty of moisture.

10.3 Low-Pressure Systems

The meeting of these contrasting air masses at the polar front leads to the development of mid-latitude weather systems. The generation of depressions – a process known as **cyclogenesis** – has been studied for a long time. The early work of Jakob Bjerknes and his group of scientists in Norway in the 1910s and 1920s (see Chapter 2) generated the so-called **Norwegian model**, which states that depressions form from waves that develop on the polar front by the steps shown in Figure 10.4. In its unperturbed state, the polar front is a long trough of low pressure with very strong **wind shear** across it – the warm air moves in a westerly direction and the cold air moves in an easterly direction. This makes the area very susceptible to rotation. In other words, if a region of stronger ascent (lower pressure) develops on the polar front, the air rushing in to the developing low-pressure centre will quickly begin to rotate. A kink develops in the polar front, centred on the pressure minimum. The developing kink is referred to as a **frontal wave**. As the system develops, the kink in the polar front continues to grow. The central pressure continues to fall, leading to sharper pressure gradients, stronger winds and more intense rain. At this stage, the system has the familiar structure of a **warm front** being chased by a **cold front**. At the warm front, warm air rides up over cold air; at the cold front, cold air undercuts warm air. Ascending air at both fronts continues to form cloud and rain along their lengths, with most of the rain falling towards the centre.

The depression then begins to decay. The first sign of the depression's demise is the process of **occlusion**. The fronts continue to spiral around each other until the cold front starts to undercut the warm front, lifting it up off the ground, resulting in an **occluded front**. The depression now has three fronts associated with it, all meeting at the **triple point**. The

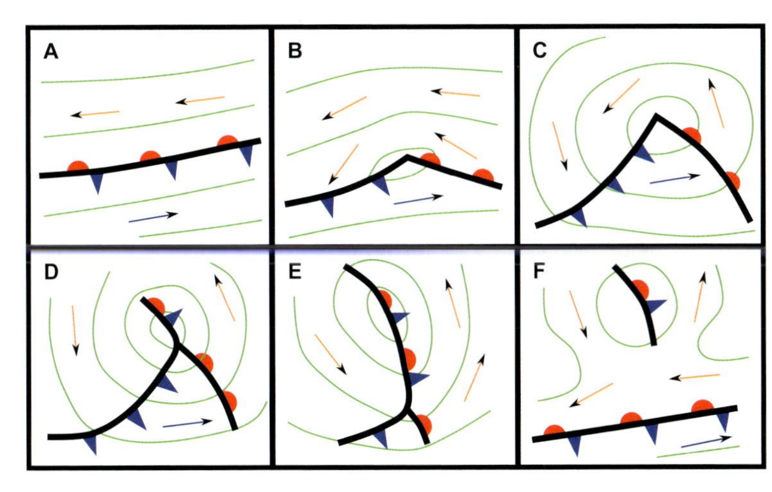

Figure 10.4 The stages of evolution of a mid-latitude depression, based on the Norwegian model of cyclogenesis. A disturbance on the polar front (**A**) leads to the development of a frontal wave (**B**). The system spirals in (**C**) until the cold front catches up with the warm front, leading to occlusion (**D**). The system then decays (**E**, **F**). (Adapted from Bjerknes and Solberg, 1922.)

winds in the system start to weaken, although cloud and rain in the centre of the system can still persist for a few more days. The total lifetime of a depression, from frontal wave to occluded system, lasts from a few days to about a week and, in this time, it could easily complete a journey across the North Atlantic. A satellite image of a mature depression is seen in Figure 10.5.

So what conditions favour cyclogenesis on the polar front? The answer to this is complicated, as there are a great many factors to consider, although there are perhaps two main ingredients. To initiate cyclogenesis, we need the ascent on the polar front to be enhanced and maintained. For this to happen, we require a convergent flow at the surface to continually push air into the ascending column from below and a divergent flow aloft to pull air out of the column from above.

Figure 10.5 A mature mid-latitude depression off the north coast of Scotland on 31 January 2008. The occluded front hooks round sharply into the centre of the depression; the trailing cold front stretches along the entire coast of mainland Europe. (Image: NEODAAS/University of Dundee.)

The pull from above comes from the jet stream. The upper-level flow, in which the jet sits, meanders north and south. As it does so, air in the jet is forced to change speed. As the flow sweeps south towards an upper-level trough, the winds speed up, leading to convergence; as the flow leaves the trough, the winds slow down again, leading to divergence (Fig. 10.6). Depressions form most readily in these divergent regions ahead of upper-level troughs. An area where conditions are perfect for depression formation is often called a **development site**. One development site can spawn whole chains of depressions on the polar front, if conditions are favourable; if the cold front from a decaying depression still has a strong enough temperature gradient, then the next depression can form in the wake of the previous one. In some instances, a whole chain of depressions in various stages of maturity can be seen to stretch across the Atlantic along the polar front.

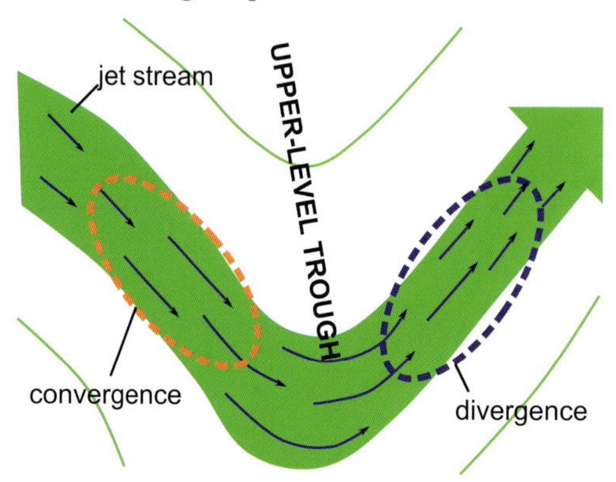

Figure 10.6 An upper-level trough is marked by an Equatorward excursion of winds in the jet stream. Winds slow down as they pile up into this trough, but accelerate as they exit it. This leads to an area of divergence aloft ahead of an upper-level trough that favours cyclogenesis (blue dashed circle), with an area of convergence behind (orange dashed circle).

Cyclogenesis occurs more favourably in areas of the world where sharp contrasts between opposing air masses are frequent. In the North Atlantic, prime conditions for storm formation occur off the east coast of Canada. Here, cold continental air from North America meets tropical maritime air, heated from below by the warm ocean. Most of the depressions that make landfall in western Europe start their lives here before tracking east-north-eastwards across the Atlantic. Usually, on making landfall, depressions begin to dissipate as their moisture supply is removed. The paths frequented by depressions are referred to as **storm tracks** (Fig. 10.7). In the northern hemisphere, storm tracks run across both the North Atlantic and North Pacific. In the southern hemisphere, storms are free to process around the Southern Ocean without ever making landfall, generating a single storm track that extends all the way around the world.

The tracks of depressions are greatly affected by the presence of regions of high pressure. Indeed, a well-established anticyclone can present a barrier to depressions, often deflecting them off course from the normal storm track. Such a disruptive anticyclone is referred to as a **blocking anticyclone**. Temporary changes to the storm track can markedly change the weather in a location. One way of quantifying the position of the North Atlantic storm track is to look at the **North Atlantic Oscillation index**. The index is determined by comparing the mean-sea-level pressure in the Azores and Reykjavik in Iceland. If the North Atlantic Oscillation index is positive, the pressure gradient is steeper than normal; if it is negative, the gradient is shallower (or indeed reversed). The positive

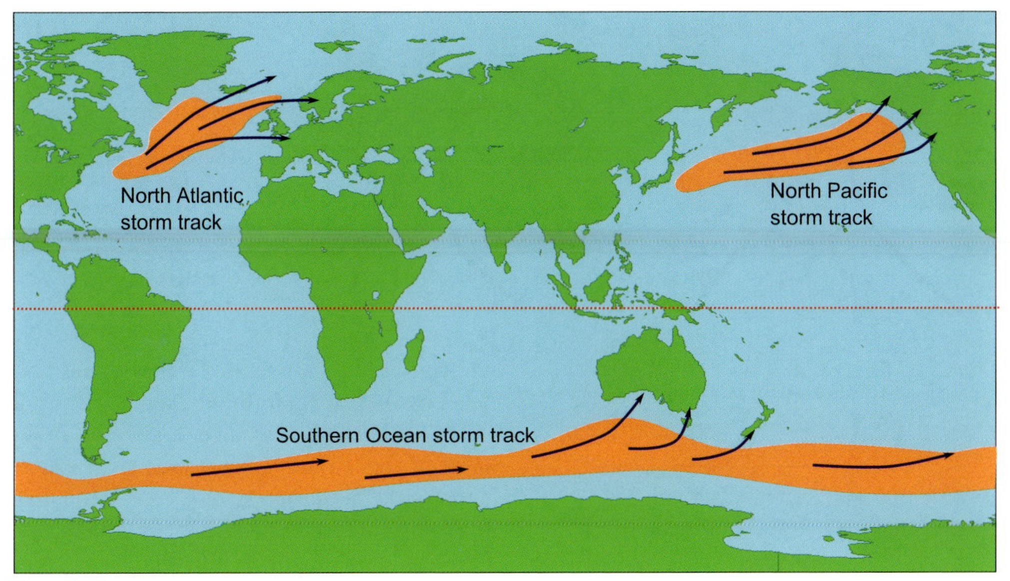

Figure 10.7 The locations of storm tracks. Northern hemisphere storms form and track east-north-eastward across the oceans; southern hemisphere storms track around the Southern Ocean.

phase indicates higher pressure to the south, hence storms track further north bringing warmer, wetter conditions to northern Europe and drier conditions to southern Europe. In the negative phase, the converse is true, with the storms tracking further south.

10.4 Fronts and Conveyor Belts

Once a depression has formed, however, it requires more than just convergence below and divergence above to keep it alive. A great deal of the energy used to develop the depression comes from the latent heat released in the creation of clouds and rain. Without a constant stock of warm, moist air, the depression would quickly begin to decay.

The transport of heat and moisture around the depression is effected by a complex series of interwoven streams of air (Fig. 10.8). Indeed, the flow patterns within the depression are far more complicated than a typical two-dimensional depiction on a weather map may suggest. For a start, the fronts within a depression are not just surface features, but extend all the way to the tropopause (although their temperature gradients can be very weak in the upper troposphere). They are also inclined to the Earth's surface at a very shallow angle – a warm front forms at an angle of less than 1° to the surface; a cold front forms at an angle of 1° to 2°. This means that ascending air at the fronts does not ascend vertically, but follows a so-called slantwise path. Therefore, cloud that forms on the fronts is typically stratiform as opposed to convective.

An example of weather observed during the passage of a depression is shown in Figure 10.9. As a depression approaches, the first

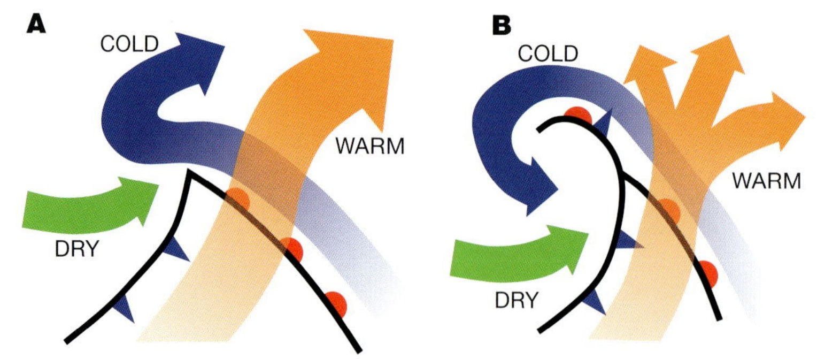

Figure 10.8 The flow within a depression consists of warm, cold and dry conveyor belts, typical paths of which are shown here for (**A**) a developing depression and (**B**) a mature, intense depression. The intensity of the colour on the warm and cold conveyor belts indicates increasing height.

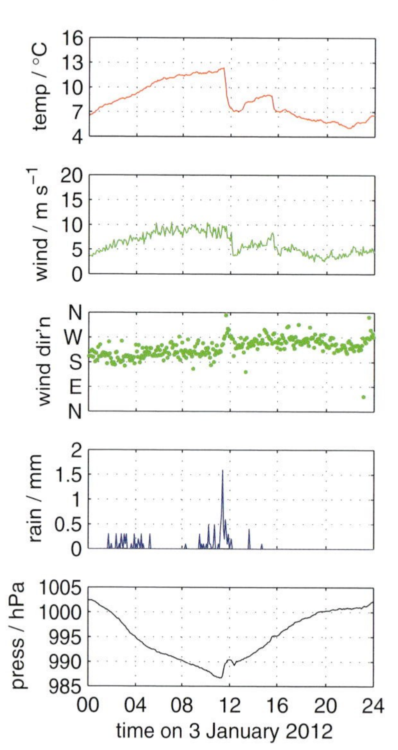

Figure 10.9 Weather records during the passage of a low-pressure system over Reading, UK, on 3 January 2012. The weaker warm front passed over at about 03:00, indicated by a period of light rain and a slight veer in the wind. The cold front passed over at 12:00, bringing a sudden drop in temperature and wind speed, veering of the wind, plenty of rainfall and a dip in the pressure. (Data: Department of Meteorology.)

indication of its arrival is the very uppermost reaches of its warm front, marked by cirrus ice clouds that thicken to form a smooth, milky layer of cirrostratus. Cirrostratus gives way to altostratus, a thicker layer of ice cloud that may obscure the Sun. As the cloud base falls below the melting layer, altostratus gives way to mixed-phase nimbostratus. Nimbostratus lacks the vertical extent to produce spectacular shows of thunder and lightning, instead providing long periods of persistent rain as the surface front passes over. As its gradient is so shallow, the topmost reaches of a warm front can precede the surface feature by several hundred kilometres, giving plenty of warning of a change in the weather (Fig. 10.10).

The warm wedge of air between the warm front and the cold front behind it is called the **warm sector**. This is the primary source of heat and moisture that drives the depression. The passage of the warm front brings higher temperatures and also a shift in the wind – from ahead of the front to behind the front, the wind will veer. Weather in the warm sector often consists of small pockets of cumulus convection that are sometimes strong enough to form showers of rain. The warm sector is followed by the cold front. Clouds forming where the cold front meets the ground tend

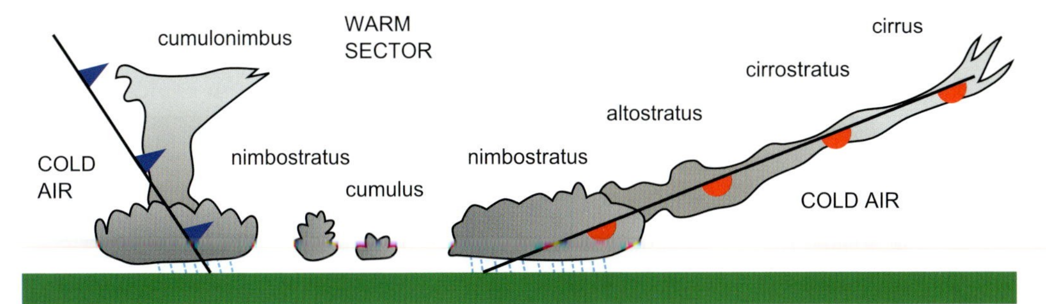

Figure 10.10 Cross-section through the warm front, warm sector and cold front of a mid-latitude depression, with the associated cloud types indicated.

to be a combination of nimbostratus and tall convection, with cumulus clouds possibly even giving way to a cumulonimbus or two. Again, as the cold front passes over, the wind will veer. Behind the cold front, the air is typically much drier, which leads to clearing skies, but colder weather can certainly follow. Cloud that forms on occluded fronts is usually mostly deep nimbostratus, which brings further persistent rain, which can be heavy in more intense systems.

It turns out that the ascent of air along both warm and cold fronts is inextricably linked. At the heart of the depression are three interacting air streams flowing through the system: the dry conveyor belt (see Fig. 10.8), the warm conveyor belt and the cold conveyor belt. The **warm conveyor belt** runs in the warm sector ahead of the advancing cold front, rising continuously and providing the centre of the system with a constant source of warm, moist air. This air acts to maintain the rain clouds associated with the cold front, and provides a constant stream of moisture to replenish the higher-level clouds found towards the top of the warm front. The **cold conveyor belt** runs along parallel to the warm front in the colder air, spiralling into the centre of the system. It provides the centre of the system with moisture and cloud and sometimes forms a hooked, comma-shaped mass of cloud referred to as a **cloud head** to the west side of the centre of the system. The **dry conveyor belt** feeds dry air behind the cold front and can sometimes become wrapped into the middle of the depression, creating a cloudless region behind the cloud head called a **dry intrusion**.

These constant flows of air provide the depression with the energy and moisture required to produce clouds and rain throughout its lifetime. Based on this model it is easy to see how the process of occlusion leads to the eventual demise of the depression; as the cold front undercuts the warm front, the complex warm conveyor belt is disrupted, which cuts off the system's supply of moisture and results in eventual dissipation.

10.5 When Storms Become Severe

Naturally, no depression has a life cycle that exactly follows that of the Norwegian model. Many depressions do indeed contain the key features of a warm front and a cold front, eventually leading to occlusion, but quite often one of the fronts will be substantially stronger than the other. Also, of course, not all depressions provide equal amounts of cloud and rain – some depressions may form and quickly fizzle out, while some become large, well-developed systems that persist for a week or more.

Even in the temperate mid-latitudes, however, it is possible for extremely rapid cyclogenesis to create a very severe storm if conditions are perfect at both the surface and the tropopause. One measure of the severity of the storm is the rate at which its central pressure falls. Air pressure at the centre of severe storms can fall by 20 to 30 hPa in 24 hours. If the pressure of a developing depression falls by more than 24 hPa in 24 hours, it is often referred to as a **bomb**. These typically are surrounded by a steep pressure gradient, implying high wind speeds. Rapid cloud development can lead to very heavy rain with the potential to cause localised flooding.

Severe storms can also bring damaging winds in the form of **sting jets**. These are generated by evaporative cooling within the cloud head. Evaporating ice particles create a mass of colder air within the cloud, which becomes denser than the surrounding air. Ice continues to evaporate into this air, further cooling it until it descends rapidly to the surface as a blast of wind. Sting jets can provide the strongest and most damaging surface winds found in severe mid-latitude depressions.

We have already seen that the formation of low-pressure systems is favoured where contrasting air masses frequently meet. However, not all of these move eastwards in the storm tracks. **Polar lows** form at very high latitudes over the Arctic, where cold air off land or sea ice meets air over the warmer (although still very cold) ocean. Polar lows do not contain weather fronts like most mid-latitude depressions, but take the form of a small system of convective cloud (Fig. 10.11). Once formed, they track southwards over the ocean, picking up moisture as they go. Some just bring cloud and light snow, although others can develop into severe weather systems, bringing vast amounts of snowfall and potentially causing widespread disruption.

Opposite, Figure 10.11 A polar low over the snow-covered coast of Norway on 6 April 2007. (Image: NEODAAS/University of Dundee.)

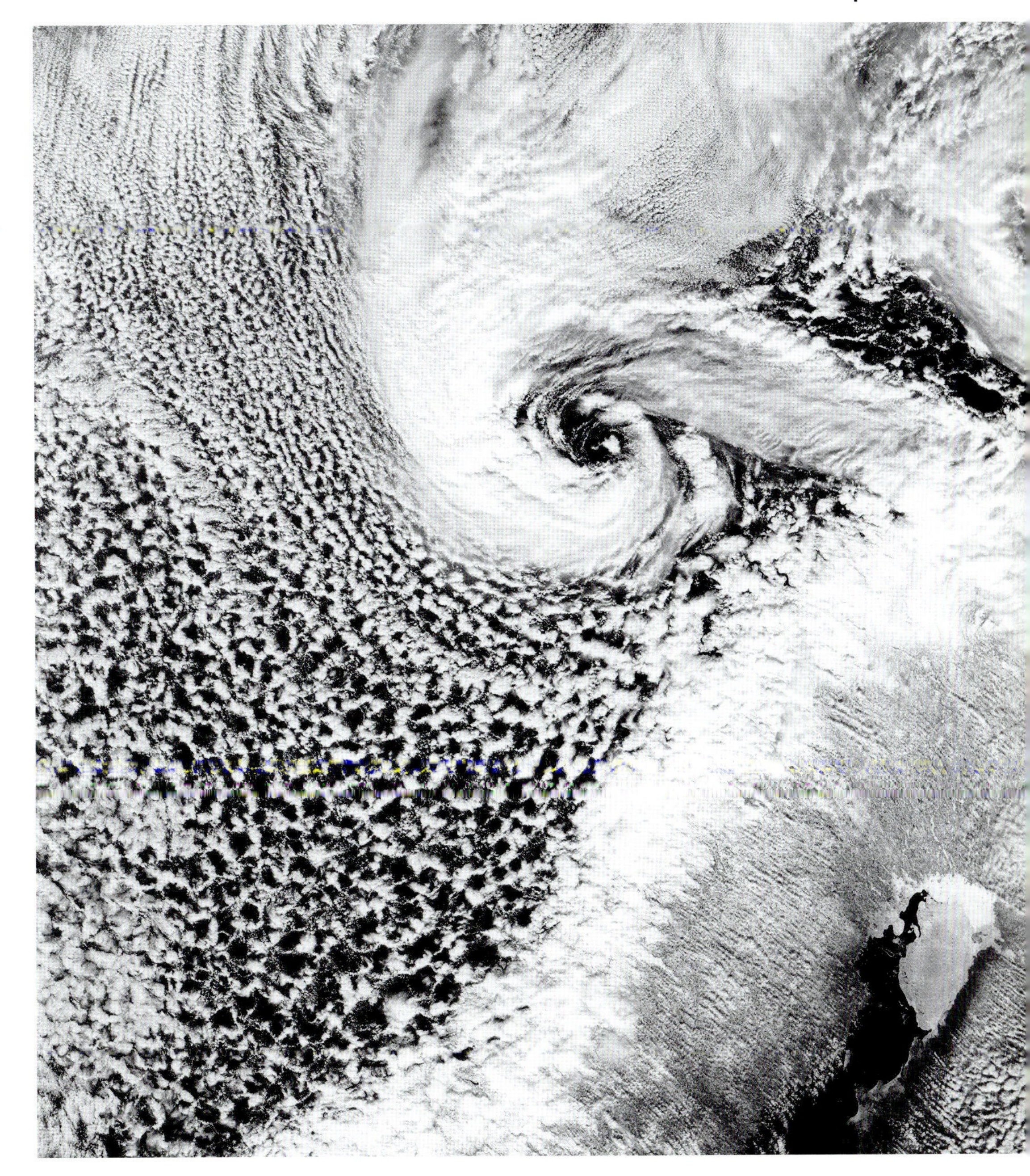

11 Weather in the Tropics

The weather found in the **tropics** is very different to mid-latitude weather. Perhaps the most significant difference between the mid-latitudes and the tropics is the distribution of temperature. Firstly, the tropics are much warmer, implying that the atmosphere contains a great deal of energy and can hold much more moisture. Secondly, solar heating varies little over the course of the year in the tropics, leading to only minor temperature swings between the seasons – indeed, the swings of temperature between day and night are usually much greater than any variability brought about by the changing seasons. This uniform heating leads to an absence of strong north–south temperature gradients across the tropics, meaning that depressions cannot form in the same way as they do in the mid-latitudes. However, contrasts in temperature can still develop between ocean and land surfaces, and these contrasts can generate weather systems. The mechanisms by which low-pressure systems form are different, but still have the potential to generate storms as damaging as mid-latitude depressions – sometimes, even more so.

11.1 The Easterly Flow

The meteorological definition of the tropics is the part of the Earth that is encompassed by the Hadley cells, the vast circulations of air described in Chapter 9 that begin the process of transporting energy polewards. These are driven by the small differences in incoming solar radiation across the tropics. Where this radiation is strongest, increased warming of the atmosphere causes air to ascend, creating a band of low pressure. The ascent continues all the way to the tropopause, where the flow turns poleward to the north and south, before descending in the vast, cloud-free areas of sub-tropical high pressure. The surface winds, commonly referred to as the trade winds, then blow back from these sub-tropical high-pressure regions towards the Equator. The Coriolis effect deflects the trade winds to be north-easterly in the northern hemisphere and south-easterly in the southern hemisphere. In contrast to the flow in the mid-latitudes, which is generally westerly throughout the troposphere, flow in the tropics (at least, near the surface) is generally easterly.

The heart of tropical weather is the line of convergence where the trade winds from each hemisphere meet, referred to as the **Inter-tropical Convergence Zone**, or ITCZ (see Fig. 11.1). This has certain similarities with the polar front in the mid-latitudes; it is a region of ascent where the formation of low-pressure systems is favourable. Indeed, most tropical depressions start their lives in the ITCZ. But there is a fundamental difference between the polar front and the ITCZ: the absence of temperature gradients. The polar front is a zone where warm air and cold air meet, creating a strong temperature gradient; however, the converging air in the centre of the Hadley cells is warm on both sides. This means that tropical

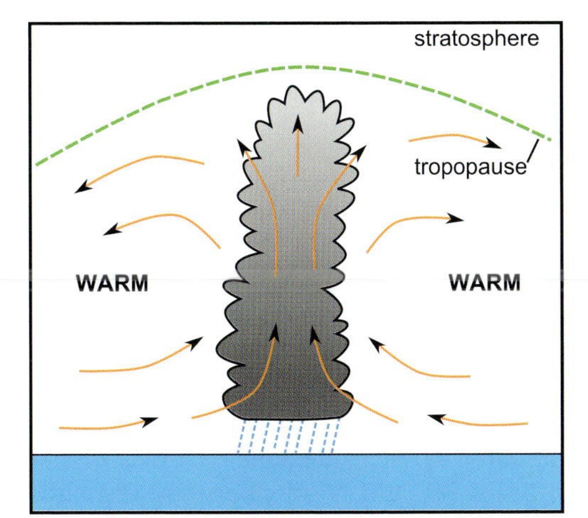

Figure 11.1 Cross-section through the Intertropical Convergence Zone. The flow of warm air from either side leads to the formation of deep convective clouds.

depressions must be very different in structure to their mid-latitude counterparts. Indeed, tropical depressions are not marked by spiralling weather fronts, but organised patches of convection and ascent that initiate cyclonically spiralling winds. However, pressure at the centre of most tropical depressions is only a few hectopascals lower than the surroundings on account of the weakness of the Coriolis effect in the tropics (the size of the effect increases with latitude away from the Equator). Despite their small pressure gradients, however, tropical depressions can still pack strong winds and bring heavy rain. In extreme cases, they can grow to become tropical cyclones, or hurricanes.

11.2 Intertropical Convergence Zone and Monsoons

Like the polar front, the ITCZ is not a continuous band of low pressure and cloud, but usually takes the form of a number of patches of scattered convection and storms that are aligned in a roughly east–west direction (Fig. 11.2). The strength of the convergence at the surface can vary greatly along its length as it winds its way across both ocean and land. Its most active areas are often over land, which heats up more rapidly than the oceans during the day, leading to ascending air enhancing the convergence. The western Pacific is also a

Figure 11.2 The band of convection across the eastern Pacific Ocean and across South America is associated with the Intertropical Convergence Zone. (Image: NASA Earth Observatory/NOAA.)

hotspot for ITCZ activity, as the ocean here is warm, providing the atmosphere above with both heat and moisture. In contrast, the cooler ocean surface of the eastern Pacific leads to less vigorous ascent, causing the ITCZ to be generally less active in this region.

The activity of the ITCZ is also affected by the **Madden–Julian Oscillation**. This affects the surface pressure around the tropics, which, in turn, impacts on where the deepest convection in the ITCZ forms. The fluctuation in pressure is small (about 3 hPa), but has a marked impact on the development of tropical convection. This pressure fluctuation travels around the ITCZ from west to east, completing one lap of the Earth's tropics in 30 to 60 days. Its low-pressure phase causes a wide patch of enhanced convection followed by a patch of clear sky. The pattern is most discernible as the oscillation tracks across the Indian and West Pacific Oceans, although its pressure and wind signals travel all the way around the tropics. Its journey around the Earth is sometimes broken when it passes over areas of cooler ocean or land, where the moisture source for the convection is temporarily cut off. Sometimes the oscillation dies out completely, only to restart some time later.

The position of the ITCZ, like the polar front, is not fixed. As the seasons change, the ascending branch of the Hadley cell tries to centre itself over the part of the Earth that is receiving the maximum amount of radiation from the Sun. This causes the band of convection in the ITCZ to shift northward and southward: the passage of the ITCZ tends to define the seasons, which are not warm and cold as in the mid-latitudes, but wet and dry. Places towards the edge of the tropics typically have one wet season per year; those near the Equator have two as the ITCZ passes over in each direction. The dry and wet seasons are also marked by reversals in prevailing wind direction. This dramatic transition from dry conditions to wet conditions only occurs in the tropics. In mid-latitude locations, rainfall tends to be fairly evenly distributed throughout the year; in the tropics, most of the rain falls in one part of the year. The reversal of wind direction brought about by the passage of the ITCZ is referred to as a **monsoon**.

The speed at which the ITCZ passes over the surface is dependent on the nature of that surface. The surface of the ocean responds fairly slowly to changes in air temperature above. In contrast, the land surface responds much more quickly. In other words, when the monsoon makes landfall, it can travel much faster and quickly sweep across large areas. Hence the ITCZ can deviate a long way from the Equator in parts of the world where it crosses land. In the central Pacific, however, the slow temperature response of the ocean causes it to travel barely at all.

Monsoons affect many tropical countries. Perhaps the best known is the Indian Monsoon, although monsoon conditions also impact western Africa, south-east Asia, northern Australia and Central America. The onset of the Indian Monsoon usually occurs near 1 June and starts over south-western India. Over the course of the next month or so, it sweeps across the whole country, reversing the direction of the prevailing wind and bringing copious amounts of rainfall. During the dry season, the north-easterly trade winds sweep cooler air across India from the north. In the wet season, the direction of the wind is reversed, coming instead from the south-west (Fig. 11.3). These south-westerly winds are an

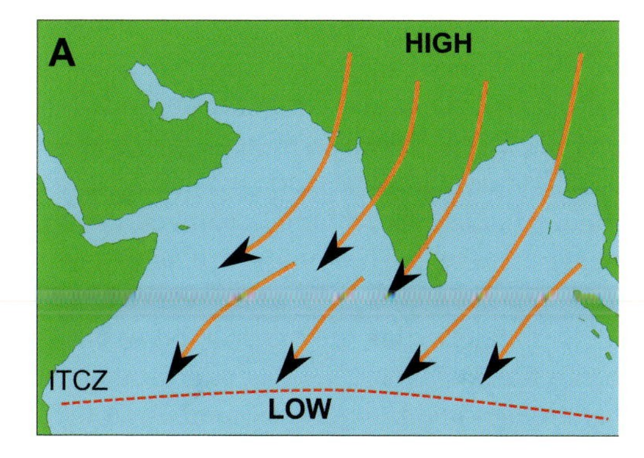

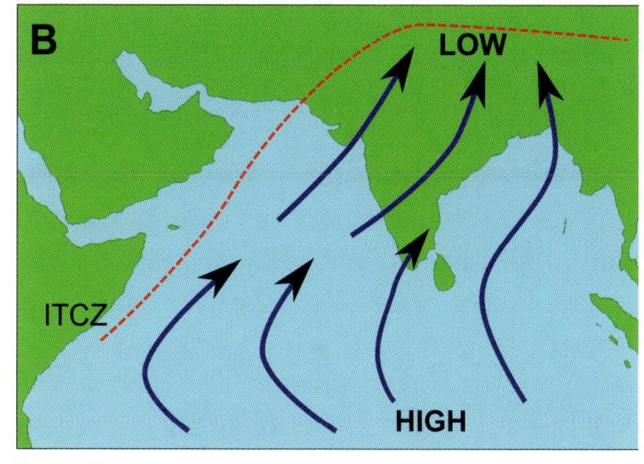

Figure 11.3 During the dry season, India is influenced by north-easterly winds coming from the Himalayas (**A**). When the wet season arrives, the winds change to the south-west, bringing moist air and wet conditions (**B**).

extension of the south-easterly trade winds of the southern hemisphere, but deflected in the opposite direction as they cross the Equator. They sweep across the surface of the Indian Ocean deep into the monsoon circulation, feeding it with heat and moisture. The retreat of the Indian Monsoon begins in September, but often takes until late October before the rains have left the whole country, after which dry conditions resume.

It should be borne in mind that the wet season is not a period of continuous rain that lasts for months on end, and the amount of rain brought by a rainy season is by no means constant. The patchiness of the convective clouds that form on the ITCZ can lead to a great deal of variability in rainfall accumulation in an area between individual wet seasons. Farming in the tropics is very dependent on the arrival of the monsoon to provide irrigation for crops, yet very sensitive to the amount of rain that falls. An excess of rain can lead to localised flooding; a lack of rainfall can lead to a crippling drought – both potentially very damaging to the economies of affected countries.

11.3 Tropical Depressions and Hurricanes

So how do depressions form in the tropics in absence of temperature gradients? As in the mid-latitudes, there are a number of factors that can influence tropical cyclogenesis, but again, the simple answer returns to the same two ingredients: convergent flow at the surface and divergent flow aloft. In other words, we need a patch of ITCZ where the convergence is enhanced to line up with either a meander in a jet stream or a divergent patch in the upper-level flow.

Jet streams do exist in the tropics: as in the mid-latitudes, they form in regions where contrasting air masses meet. Despite the lack of large-scale temperature gradients in the tropics, local temperature contrasts can develop that are strong enough to drive a jet stream, although the gradients only usually extend a few kilometres up into the atmosphere, forming low-level jets at heights of about 3 km. Such local temperature contrasts can develop between land and sea. A prime example exists between the hot land of West Africa (with the Sahara on its northern boundary) and the cooler waters of the Gulf of Guinea, particularly during the northern summer. The contrast drives a low-level jet that travels from east to west and is referred to as the **African Easterly Jet**. Meanders in this jet, in conjunction with waves in the flow high up in the troposphere, can provide the divergence needed to set tropical cyclogenesis in motion.

Regions of upper-level divergence are often associated with patches of deep convection that slowly travel to the west. In some cases, these can become organised to form squall lines, which can bring gusty winds and heavy showers of rain (see Chapter 12). A great deal of convection in the tropics forms during the afternoon, after the atmosphere has become unstable from solar heating throughout the day. Some of these patches of convection grow to become **tropical depressions** – low-pressure centres that contain organised convection, consisting mainly of cumulus and cumulonimbus clouds that bring heavy showers of rain. In terms of pressure differences, they are shallow features, but can still produce high winds.

However, in favourable conditions, tropical depressions can strengthen to the point that they become damaging. If they can sustain wind speeds of over $16\,m\,s^{-1}$, they become known as **tropical storms**; if they can reach speeds of over $33\,m\,s^{-1}$, they become **tropical cyclones**, more commonly referred to as **hurricanes**, **typhoons** or simply cyclones. The wind speeds sustained in a hurricane are further classified into five categories according to the **Saffir–Simpson Hurricane Scale**, determined by the US National Hurricane Center. Any hurricane has the potential to do damage, but a category-five hurricane is likely to cause extensive, catastrophic damage. Category-five hurricanes pack winds of over $70\,m\,s^{-1}$ (about 150 mph) and have central pressures below 920 hPa – the most severe hurricanes can contain pressures even below 900 hPa. Tropical weather systems that make it to the status of tropical storm are usually given names. In the North Atlantic, storm names are determined by the National Hurricane Center and cycle through the alphabet each year.

Fortunately, not all tropical depressions reach hurricane status – again, a number of favourable conditions are required. For a start, hurricanes cannot form within 5° of latitude either side of the Equator, as the Coriolis effect is not large enough to set the hurricane rotating. The sea surface below must be warm enough to provide the hurricane with sufficient energy and moisture – any less than 26°C and the hurricane cannot develop. These two conditions are met in all of the tropical oceans except in the south-east Pacific, where sea surface temperatures are cooler, and in the southern Atlantic, where only one cyclone has ever been observed to form. The air contained in the developing cyclone must be warm so that it can receive enough moisture. Hurricanes also require moisture in the mid-troposphere, and a lack of wind shear throughout the troposphere, or the circulation required to generate the cyclone will not be able to start.

A fully fledged hurricane, as seen in Figure 11.4, contains a complex series of different circulation patterns within. The strongest ascent occurs in a ring around the centre of the hurricane, where deep cumulonimbus clouds fill the entire depth of the troposphere, bringing heavy rain. This region also contains the strongest winds. In the northern hemisphere, the most damaging winds in a hurricane are located to the right of its centre in the direction of travel (left in the southern hemisphere), as

Figure 11.4 Hurricane Frances on 31 August 2004 off the north-east coast of Puerto Rico. Its eye is clearly visible in the centre. (Image: NASA Earth Observatory.)

the wind speed here combines with the speed of motion of the hurricane. Bands of deep convection spiral outward from this inner ring, forming the spiral arms that make up part of the hurricane's galactic shape. However, the height of these outer storm clouds is limited by the subsiding air falling out of the vast anvil shield at the top of the hurricane. While the centre of the storm is turning violently in a cyclonic direction, this upper-level descending

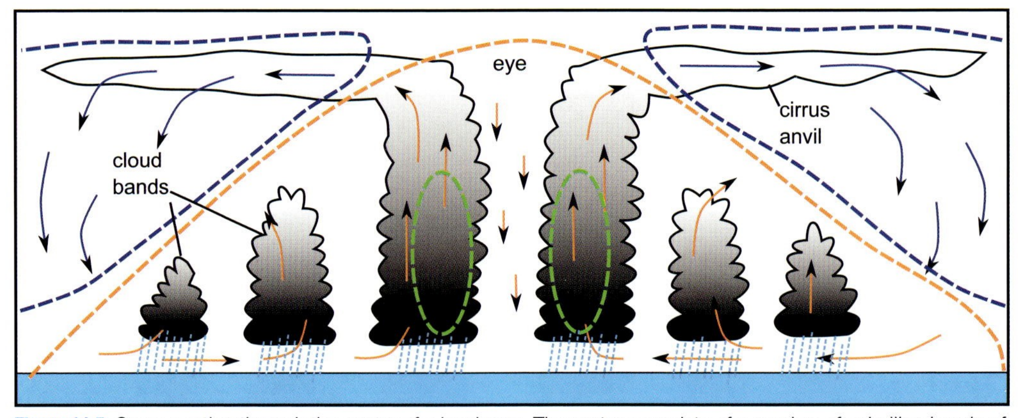

Figure 11.5 Cross-section through the centre of a hurricane. The system consists of a number of spiralling bands of clouds, with the most intense wind and rain in the eye wall (within the green lines). The centre of the flow (within the orange line) is cyclonic, while the upper-level flow away from the eye (within the blue line) is anticyclonic.

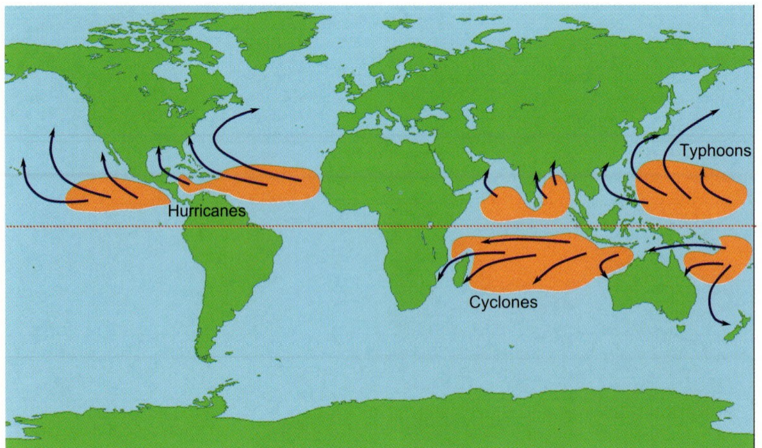

Figure 11.6 The orange regions indicate where tropical cyclones, and similar storms, typically form. Their tracks are typically to the west and away from the Equator in the tropics. If they reach the mid-latitudes, their tracks swing to the east, although they adopt the characteristics of normal mid-latitude depressions. Tropical cyclones are known as hurricanes, cyclones and typhoons in different parts of the world as shown.

air actually rotates around the system anticyclonically (Fig. 11.5). Many mature hurricanes also develop an **eye** – a patch of clear sky in the centre of the system. Amidst all the heavy winds in the surrounding clouds (known as the **eye wall**), the eye contains calm conditions with clear skies. The reason for this strange juxtaposition of weather types is a region of warm, subsiding air at the centre.

The paths taken by individual hurricanes can be erratic and tricky to predict, although the general movement of hurricanes is the same: they start off tracking west across the tropical ocean, before slowing and swinging polewards (Fig. 11.6). Here, they become entangled in the westerly propagation of the mid-latitudes and move north-eastward in the northern hemisphere (south-eastward in the southern hemisphere). Hurricanes decay when they are cut off from their energy source – the warm sea. This happens in one of two ways: either they make landfall, or they travel too far away from the Equator, where the sea becomes too cold to sustain the circulation. Consequently, even though some Atlantic hurricanes do end up affecting the weather in western Europe, there is no danger of them still being active hurricanes. Hurricanes are also resistant to crossing the Equator, as this would require them to travel through a zone with very little Coriolis effect.

Hurricanes are perhaps the most damaging of all meteorological events on account of the massive areas they can sweep over. Most of the devastation associated with hurricanes is caused by the intense winds and the heavy rain that falls near the hurricane centre. Flooding is another common impact of hurricanes, not just by heavy rain, but by **storm surges**. The low pressure in the centre of the system causes the sea level to rise with respect to the higher pressure surrounding it. This pulls a head of water into the system, which is then pushed around by the high-speed winds over the sea surface. In low-lying areas, this excess water can cause vast amounts of flooding. In the northern hemisphere, the greatest storm surges are found to the right of the hurricane's direction of travel.

11.4 El Niño, La Niña and the Southern Oscillation

One final tropical weather phenomenon that merits discussion is the **Southern Oscillation**. This is marked by changes in the circulation of air over the tropical Pacific Ocean and affects the distribution of convection on the ITCZ across the entire Pacific. We have already seen that convection over the tropical Pacific forms more favourably over the central and western Pacific, with less convection generally forming over the east. To explain this, we must turn to the ocean. In a similar way to the atmosphere, the ocean is also heated unevenly across the Earth. The temperature of the sea surface can have significant effects on the weather in a region, and it turns out that the central-to-eastern Pacific is an area where colder upwelling water reaches the surface. In other words, the sea surface here is much cooler than the rest of the equatorial Pacific. One reason for this is the effect of the trade winds. With no land to get in their way, these winds can blow unobstructed across the surface of the Pacific. As they go, they also have the effect of dragging on the ocean surface, towing water across the ocean to the west. By continuity, this water must be replaced by water pulled upwards from the ocean depths. This sets up a temperature gradient across the Pacific Ocean, with warmer waters to the west (usually centred on about 180° longitude) and cooler waters to the east, off the coast of South America. This gradient over the surface sets up an east–west circulation in the atmosphere above, enhancing ascent and convection over the central Pacific and suppressing it over the eastern Pacific. This east–west circulation is known as the **Walker circulation** (Fig. 11.7).

However, the Walker circulation is not constant, and many things can alter its position and strength. For instance, sometimes the trade winds over the Pacific become weaker. The warmer water that was being pushed to the west is now allowed to flow back into the eastern Pacific, resulting in a significant warming of the sea surface off the coast of South America. Such a warming event

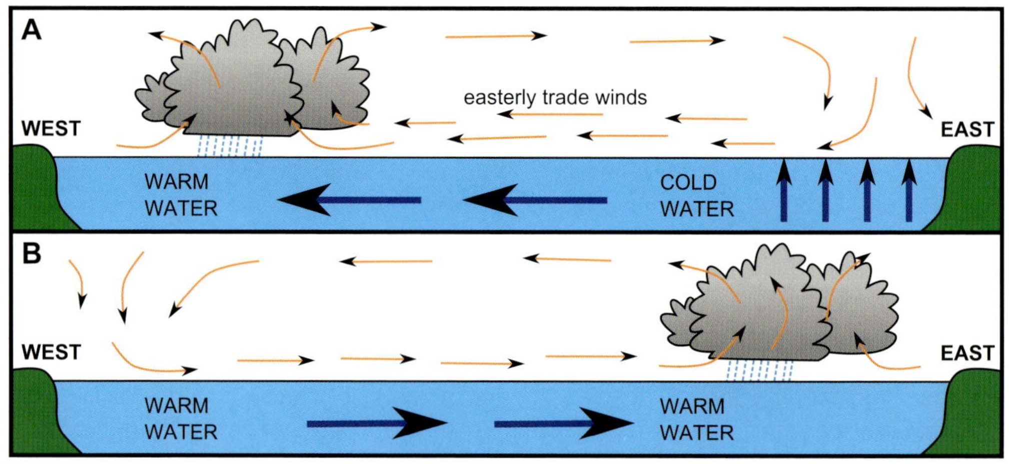

Figure 11.7 Circulation of the tropical Pacific Ocean and the atmosphere during (**A**) normal conditions and (**B**) El Niño conditions, here shown as a west–east cross-section. Under El Niño conditions, weaker trade winds lead to warm water moving eastwards and increased cloud development over the eastern Pacific.

is referred to as **El Niño**. This has dramatic consequences on the cloud patterns over the Pacific Ocean, with the patches of deep convection that were previously over the central Pacific shifting east, bringing anomalously large volumes of rainfall to the eastern margins of the Pacific.

Eventually, of course, the trade winds will regain their strength, and the warmer water will once more be pushed back towards the central Pacific. However, it is possible that the rush of warm water to the west will be pushed too far and over into the western Pacific, dragging an increased amount of cooler water from the depths into the eastern Pacific. These conditions are referred to as **La Niña**, and again affect the location of prevalent deep convection over the tropical Pacific, this time moving it west towards the western Pacific islands and bringing wetter than average conditions.

The swing between El Niño and La Niña conditions in the atmosphere, referred to as the Southern Oscillation, is not by any means regular. The transition between the two states can be fairly sudden, occurring in just a few months, and conditions can persist either way for a few months or a few years. Indeed, the oscillation can also persist in neutral conditions. An index is used to define the current state of the Southern Oscillation, which is based on a comparison of surface pressures measured at Darwin in Australia and the Pacific island of Tahiti: a negative index indicates El Niño conditions, a positive index indicates La Niña conditions.

The impacts of the Southern Oscillation are not just felt within the tropics. Such a dramatic change of atmospheric circulation over the tropical Pacific has far-reaching ramifications for weather in many other locations around the Pacific, including outside the tropics. Such distant correlations between weather patterns are known as **teleconnections**. El Niño conditions are associated not only with drought in the western Pacific and excess rainfall over Central and South America, but can also weaken the Indian Monsoon. Drier weather can also be brought to the Caribbean, with potential impacts on tropical cyclones here. El Niño can also bring warmer weather over much of the mid-latitude northern Pacific (Fig. 11.8).

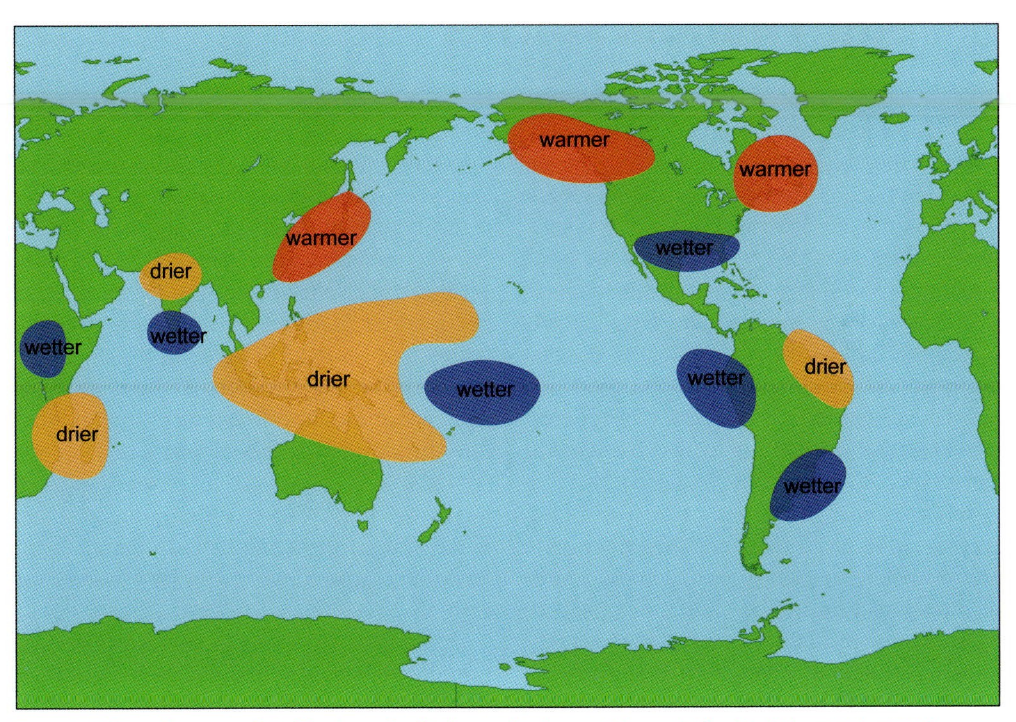

Figure 11.8 The influences of the Southern Oscillation can be far-reaching and affect both temperature and precipitation: this map shows the teleconnections associated with El Niño conditions. (Adapted from Halpert and Ropelewski, 1992, and Ropelewski and Halpert, 1987.)

12 Convective Systems, Tornadoes and Thunderstorms

Stratiform and convective clouds take very different shapes. Stratiform clouds form in layers and extend over wide areas; convective clouds are tall and extend vertically into the atmosphere. Also, stratiform clouds tend to bring long periods of persistent rain, while convective clouds bring showers. We have already seen how mid-latitude depressions contain an organised combination of stratiform and convective clouds, while tropical depressions consist primarily of convective clouds. However, convection is not just confined to weather systems. In any part of the world, convective clouds can form in an atmosphere with an unstable temperature profile and a source of moisture. Convective clouds are perhaps most often associated with a hot summer day, building in height throughout the afternoon and occasionally providing a stormy evening with hail, thunder and lightning (although they can form at any time of the day or night). Convection can take the form of individual cells, which could bring a single shower of rain, or a much larger **mesoscale convective system**, which could drop a great deal of rain and hail, produce a spectacular show of lightning, and even spawn a tornado or two.

12.1 Unstable Conditions

On a fine, clear day, the first clouds that form are often small, shallow cumulus clouds, known as **cumulus humilis**. As the surface is warmed, rising plumes of warm air called thermals pass through the lifting condensation level, causing water vapour to condense (see Chapter 8). Such small cumulus clouds are often referred to as **fair-weather cumulus**. Yet it is from fields of these small, innocuous, fair-weather clouds that thunderstorms and convective systems eventually develop. Fortunately, not every cumulus cloud grows to form a cumulonimbus thundercloud – if they did, the world would be a very stormy place. We have seen so far that, for a number of different types of weather system, development is dependent on a number of factors being favourable. The same is true for a convective system.

So how do these small clouds develop into thunderstorms? The first key ingredient is an unstable atmosphere. In Chapter 8, we saw how the difference between the temperature profile of the atmosphere (known as the **environmental lapse rate**) and both the dry and saturated adiabatic lapse rates governs the shape and size of clouds. If deep, convective clouds are to form, we need a deep, unstable layer that extends through as much of the troposphere as possible. The more unstable the layer, the more energy is available to the rising plume of air at the heart of the convective system. The second ingredient is moisture – and plenty of it. If a cumulonimbus cloud is to grow, it needs a continuous supply of moisture to be fed into the ascending air at its base. It is also favourable for the atmosphere to be moist throughout its depth, so that the growing cloud does not evaporate too quickly around its edges.

The generation of a deep, unstable layer can happen in two ways. The first is surface heating, which is how afternoon thunderstorms form on a hot summer's day in the mid-latitudes, or during a wet-season afternoon in the tropics. As the surface is strongly heated, the temperature of the lower atmosphere can become several degrees warmer than the layers above. The same can happen if a cool air mass passes over a warmer surface. Alternatively, cooling of the atmosphere aloft can also destabilise the atmosphere.

12.2 Cumulonimbus, Thunder and Lightning

There are many different types of convective system, each with a different typical lifetime, size and intensity. The principles of their formation and dissipation, however, tend to follow the same life cycle. The early stages of cumulonimbus cloud development are marked by the rapid vertical growth of cumulus clouds. In an unstable troposphere, the cumulus clouds can grow very tall very quickly as the ascending plume extends higher and higher into the troposphere (Fig. 12.1). The condensation of water droplets releases latent heat, giving the plume even more energy to ascend. Tall cumulus clouds are referred to as **cumulus congestus**. At the heart of the growing cloud is a column of ascending

Figure 12.1 Small cumulus humilis (**A**) can quickly grow into much larger cumulus congestus (**B**), and eventually become a full-grown cumulonimbus thundercloud (**C**). (Photos: Wagner Nogueira Neto/Ana Carolina Lacorte de Assis (A), Jon Shonk (B), Robin Hogan (C).)

air, referred to as an **updraught**. As air tumbles out of the updraught into the surrounding air, it creates billows and eddies that give cumulus congestus clouds their cauliflower-like shape. Water droplets near the cloud edge evaporate into the surrounding air, increasing the moisture around the cloud.

The transition of the cloud from cumulus congestus to cumulonimbus is marked by the formation of ice crystals at the top of the cloud, giving it a fuzzy appearance. Once this glaciation begins, even more latent heat is released, providing the cloud top with a boost of energy and allowing it to rapidly grow further into the troposphere. If the updraught has sufficient strength, the cloud could extend all the way to the tropopause. As it encounters this stable layer, it spreads out to form a cap of cirrus cloud called an **anvil cloud** (Fig. 12.2). Often a small dome of cloud forms at the top of the updraught as it tries to penetrate the stratosphere, referred to as a region of **convective overshoot**. At its greatest size, a cumulonimbus thundercloud typically contains between 1×10^8 kg and 1×10^9 kg of ice and liquid water.

By now, there is plenty of energy within the cloud to form precipitation. The precipitation falling from the cloud is a combination of raindrops from near cloud base and melted ice crystals and snowflakes from higher up in the

Figure 12.2 Convective clouds of various sizes over Spain and Portugal. The large, fuzzy clouds are the cirrus anvils of cumulonimbus clouds. (Image: NEODAAS/ University of Dundee.)

cloud. The updraught at the heart of the cloud is now accompanied by a **downdraught**, creating circulation within the cloud. The downdraught is driven by drag on the air by falling precipitation – for this reason, the heaviest rain tends to fall out of the cloud in unison with the downdraught. Air in the downdraught is usually cold, mid-tropospheric air that descends all the way down to the ground, creating a spreading pool of cold air. The advancing edge of this cold pool is referred to as a **gust front**, and usually carries a sharp burst of wind. The vigorous circulation within the cumulonimbus can create undulations on the base of the cloud, known as **mammatus** (Fig. 12.3).

The thunderstorm is now in its mature phase. Strengthening circulations can create hailstones as ice and liquid particles are transported around the cloud. During many trips up and down the cloud, an ice crystal can alternately gain layers of watery ice from near the cloud base and layers of more solid ice from the cloud top. It continues to circulate around the cloud until it becomes too heavy to be supported by the updraught. This means that the size of hail is indicative of the strength of the circulations within. Typical hailstone sizes vary from a few millimetres to a centimetre or two – although in severe convective storms, hail can become much larger in size. The largest hailstone ever recorded was over 18 cm in diameter.

Figure 12.3 Mammatus over Harpenden, UK on 02 April 2006. Mammatus are udder-shaped undulations found on the underside of a thundercloud, brought about by turbulence at cloud base. (Photo: Andy Barrett.)

A storm capable of delivering hail to the ground is likely to also provide a spectacle of thunder and lightning. Lightning is a spark of electricity caused by the separation of charged particles within the thundercloud. Ice particles form preferentially at the top of the cloud and liquid droplets at the bottom. As they travel around in the cloud, they build up a static charge. The solid ice crystals take on a positive charge; the liquid droplets and hailstones take on a negative charge. But the circulations do not distribute the ice and liquid particles evenly throughout the cloud; often, different types of particle amass in different regions of the cloud. The charge separation grows as the particles collect, until it is neutralised by the spark that is lightning. This occurs when the potential difference exceeds a certain level, allowing the air, normally a good electrical insulator, to suddenly ionise and conduct electricity. The air within a stroke of lightning is rapidly heated to over 30,000°C, creating rapid expansion and a shock wave. This shock wave is what we perceive as thunder. From a nearby lightning strike, thunder is heard as a sudden crack; from further away, it is heard as a rumble as the sound waves reflect off solid objects and the energy in the shock wave dissipates. We hear the thunder after we see the lightning, on account of the difference between the speed of light and the speed of

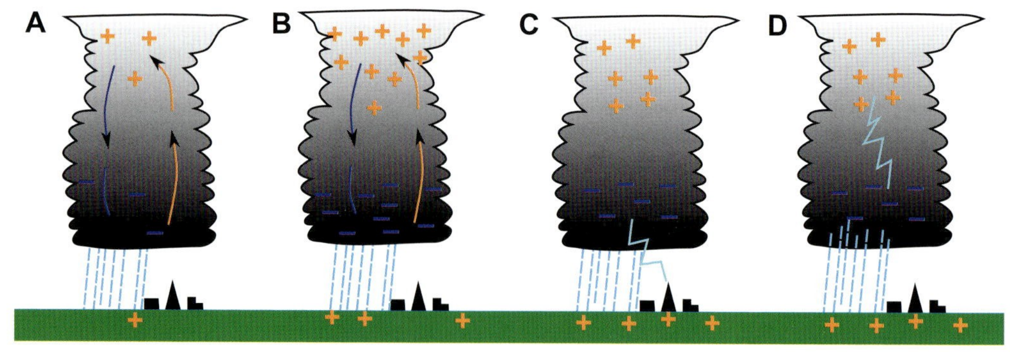

Figure 12.4 Lightning is caused by the separation of positively and negatively charged particles within a thundercloud. The circulations within the cloud carry positively charged ice crystals to the top and negatively charged liquid droplets to the bottom (**A**, **B**). This creates a potential difference, which is neutralised by a spark of lightning, either from the cloud to the ground (**C**) or within the cloud (**D**).

sound – providing the handy rule that a three-second delay between lightning and thunder implies the strike happened about 1 km away.

Lightning can travel between patches of separated charge within the cloud (referred to as **sheet lightning**) or can pass between the cloud and the ground (usually referred to as **forked lightning** on account of its branching shape). Most cloud-to-ground lightning is **negative lightning**, occurring when the cloud base adopts a negative charge, inducing a positive charge on the surface and causing current to flow upwards, as seen in Figure 12.4.. The negatively charged cloud base sends out a **leader stroke** to the surface. This creates a path of ionised air through which the much brighter **return stroke** can travel, neutralising the charge difference. It may take several strokes for the charge neutralisation to be complete – the single flash we perceive as lightning is often a number of rapid sparks, one after another. On rarer occasions, cloud-to-ground lightning can be positive. Under such conditions, positive charge accumulates at the cloud base, inducing a negative charge on the surface, and the current in the resulting spark flows in the opposite direction. **Positive lightning**, accounting for about 10% of all strikes, contains a much larger current and has the potential to be much more damaging, but is usually only found in the most severe thunderstorms.

Eventually, the thunderstorm starts to dissipate. Most individual cumulonimbus thunderclouds – often referred to as **single-cell storms** – are short-lived, with lifetimes of about 30 minutes. The usual cause of dissipation is the cutting off of the thundercloud's supply of warm, moist air. A strengthening downdraught can quickly stifle the updraught at the centre of the cloud, or disturb the flow of moisture into the cloud from below.

12.3 Organised Convective Systems

A single-cell cumulonimbus rarely has the strength to cause any damage. However, a number of different types of convective system exist where single-cell thunderclouds combine to form **multi-cell storm** systems that are much more robust and can deliver heavier rain, larger hail and more intense thunder and lightning. If they grow strong enough, multi-cell convective systems have the potential to cause widespread damage.

A multi-cell cluster of thunderstorms is a common occurrence – in most locations, thunderstorms tend to travel in groups that seem to persist for much longer than the half-hour lifetime of a single-cell thundercloud. In reality, these groups are made up of a number of thunderstorms in various stages of their lives, with new ones constantly forming in the wake of decaying ones. If the atmosphere remains unstable, thunderstorms can form in this way for long periods of time – certainly a few hours.

The weakness of a cumulonimbus is the tendency for its downdraught to stifle its energy source as precipitation falls into the updraught. However, if the updraught and downdraught align themselves within the cloud in such a way that the downdraught never stifles the updraught, a thunderstorm can become much longer lived. This often happens when the thunderstorms form in an area of strong wind shear. The wind shear pushes the updraught away from the vertical, creating a cloud that leans backwards with respect to its direction of travel. The compensating downdraught can then form underneath the updraught, with the precipitation falling out of the updraught, but never interfering with the updraught or the supply of moisture sourced from ahead of the cloud.

Convection can also organise itself with the large-scale flow. Often, the presence of a line of upper-level divergence is enough to trigger a sharp line of convection, referred to as a **squall line** (Fig. 12.5). As with most convection, these can occur in most parts of the world. An approaching squall line appears as a line of deep, convective cloud. The first indication of its arrival is the gust front, marked by a sudden increase in wind speed, a change in wind direction and a drop in temperature. The heaviest rain falls out of the back of the cumulonimbus in the squall line, where the downdraught forms. The heavy rain is often short-lived, although squall lines can develop regions of stratiform cloud behind them that bring periods of light rain.

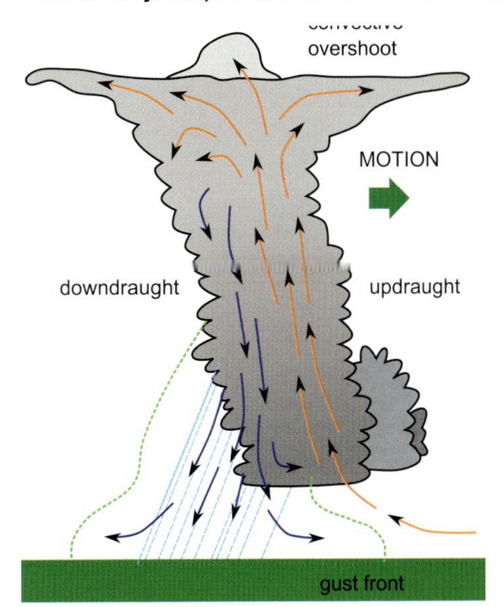

Figure 12.5 A squall line is an organised band of convection, here seen in cross-section. The slope of the updraught and downdraught prevents rain falling out of the cloud into the updraught and stifling it. The cold air falling out of the cloud with the rain spreads out to form a gust front.

12.4 Supercells and Tornadoes

The most severe type of thunderstorm is the **supercell**. Of the convective systems that form around the world, very few make it all the way to supercell status. Supercell thunderstorms form when there is both linear and rotational wind shear with height. This rotation causes the updraught and downdraught in the system to spiral around one another, creating a **mesocyclone**; the linear shear ensures that the updraught and downdraught do not meet. The rotation flings precipitation out of the updraught, further preventing any stifling effects. The passage of a supercell is likely to bring huge amounts of rain, hail, thunder, lightning and damaging winds – and even **tornadoes**.

Supercells form in a number of different locations, but perhaps the best-known location for the formation of supercells is in the central and eastern USA (referred to as **Tornado Alley**)(Fig. 12.6). From April to June, the conditions are right for the formation of supercells and tornadoes, enticing storm chasers from all over the world. Hot, dry air from the land to the west overruns warm, moist air from the Gulf of Mexico. Surface warming creates perfect conditions for deep convection, but with a strong capping inversion in the hot, dry air. As the convection builds, the developing thermals will eventually become strong enough to break through this inversion, allowing tall, deep convection to develop rapidly. A few of these massive cumulonimbus clouds may develop into supercells.

As the two masses of air approach from different directions, there must also be a fair amount of rotational wind shear, making the area a prime location for the formation of tornadoes. The exact process by which tornadoes form remains uncertain, although

Figure 12.6 Supercell activity over Tornado Alley: (**A**) a wall cloud near Greensburg, Kansas, on 09 June 2009; (**B**) a wedge tornado near Attica, Kansas, on 12 May 2004; (**C**) a rope tornado near Fairview, Oklahoma, on 24 May 2011. (Photos: Robin Tanamachi (A, B); Gerard Devine (C).)

current theories suggest that they originate as a horizontally aligned, rotating column of air. This forms in the boundary region between the warm, moist air below and the hot, dry air above. The circulation within the cloud twists and distorts this rotating column so that it ends up aligned vertically within the supercell. Air rises up this spinning column, providing an updraught into the cloud. Where the column meets the cloud base, there is typically a mass of cloud extending out of the cloud base. This is called a **wall cloud** and encloses the area of the cloud where there is most horizontal rotation. It is from within this wall cloud that a tornado is most likely to form (Fig. 12.7).

Tornadoes are narrow, rapidly rotating vortices that are renowned as some of the most devastating natural phenomena on Earth on account of their strong winds. A severe tornadoe can be a few kilometres across, contain winds of 130 m s^{-1} (300 mph) and have an air pressure in its centre that is way below any pressure attained by a mid-latitude depression or tropical cyclone. During its life, which could be as long as an hour, it can travel a long distance and cause plenty of destruction. Most damage from a tornado is brought about by the high wind speeds; few buildings are built to withstand such an onslaught.

Of course, tornado formation is not just limited to Tornado Alley – if the conditions are correct, they have the potential to form anywhere in the world. Indeed, some countries report a larger number of tornadoes per unit area than the USA. As with many weather features, tornadoes form on a wide range of scales (most are between 100 m and 500 m across) and most have lifetimes of the order of minutes rather than hours. Some tornadoes sweep across the surface of the ocean, where they are known as **waterspouts**. Indeed, many tornadoes never reach the ground. Rotation in a storm cloud can sometimes generate a **funnel cloud** – a small vortex of condensed water vapour extending out of the cloud base. Funnel clouds tend to occur in mid-latitude convective systems where there is strong wind shear or twisting in the air flow. The intensity of a tornado is rated on the **Enhanced Fujita–Pearson Scale**. There are six categories of tornado running from EF0 to EF5. The EF rating is made qualitatively by the amount of damage caused by the tornado (quantitative measurement of tornadoes is very difficult, as they are so localised and it is often impossible to place measuring devices in their track). A small number of EF5 tornadoes occur each year in the central and eastern USA.

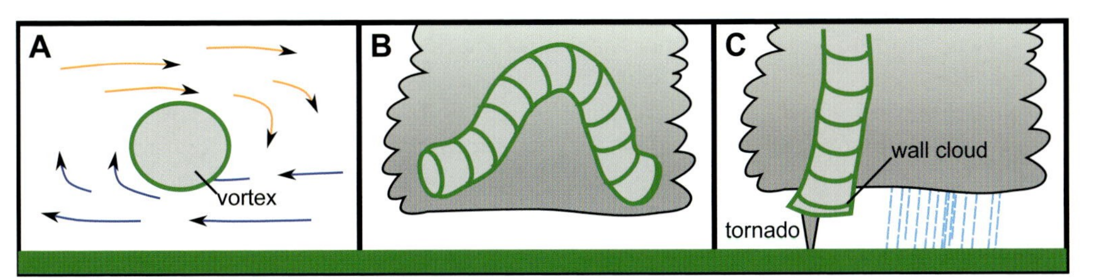

Figure 12.7 Tornado formation. (**A**) Strong wind shear sets a horizontal tube of air (shown here in cross-section) rotating. (**B**) This tube is then distorted within the developing supercell, eventually rotated vertically (**C**) and acting as the inflow to the storm. This region is surrounded by a wall cloud and may contain tornadoes.

13 Local Weather Effects

Weather systems form on a wide range of scales – mid-latitude depressions and tropical cyclones that can be thousands of kilometres across; medium-sized or **mesoscale** convective systems a few hundred kilometres across; and individual thunderstorms, a kilometre or two. But weather can also occur on scales smaller even than a kilometre. We have already seen how differences in the nature of the Earth's surface can drastically impact the large-scale weather scenarios. For example, the contrast between warm, moist maritime air and cold, dry continental air across the coast of eastern Canada creates a hotspot for the formation of mid-latitude depressions; and temperature gradients between air over land and sea in the tropics can sometimes be strong enough to generate a tropical depression. The Earth's surface can have a number of interesting effects on the weather conditions, mostly on small scales.

13.1 Coastal Weather

We have already seen the effects that the sharp contrasts existing between air over land and sea can have on large-scale atmospheric motion. However, these contrasts can also impact weather on a much more local scale. Air over the sea tends to contain a great deal more moisture than air situated over land, as there is a constant supply of water vapour evaporating from the ocean surface. Also, land and sea surfaces have very different thermal properties. Land surfaces have a much lower **heat capacity** than the sea, implying that it takes less solar heating to warm the land by 1°C than it does to warm the sea by the same amount. On a sunny day, this difference in heat capacity can lead to local temperature gradients across the coast, with air above the land being heated more strongly than air over the sea. This creates a pool of warmer air over the land, which is more buoyant and begins to ascend. To compensate this, cooler air from over the sea is drawn inland to replace the ascending warm air. This leads to low-level convergence over the land and divergence over the sea, which sets up the circulation. The resulting cool wind off the sea is commonly referred to as a **sea breeze** (Fig. 13.1A).

The strength of the sea breeze circulation is driven by the difference in air temperature between the land and the sea. Because of this, the strength of the sea breeze usually peaks by mid-afternoon, when the land–sea temperature gradient is at its strongest. In the mid-latitudes, sea breezes mostly occur in summertime in calm, anticyclonic conditions, where winds are light and sunshine duration is maximised. Sea breezes also occur in the tropics, where they tend to be stronger on account of the increased solar heating.

In both the mid-latitudes and the tropics, a strong sea breeze can extend over 100 km inland. If the breeze is strong enough, this can lead to a band of cloud that travels away from the coast as the mass of cool, maritime air sweeps further inland. The sharp boundary between the sea air and the warm land air is called a **sea breeze front** and is marked by

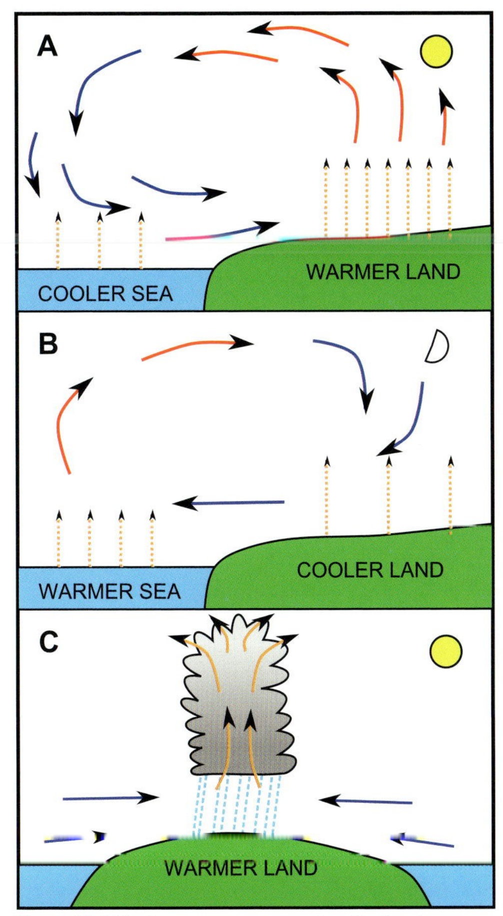

Figure 13.1 During a warm, sunny day, the land becomes warmer than the sea. This drives a circulation that causes a sea breeze (**A**). At night, the land cools more rapidly, reversing the circulation (**B**). Where sea breezes meet across a peninsula, convergence can initiate cloud formation and rain (**C**).

ascent of the warm land air. In some cases, the ascent on the sea breeze front can be enough to generate light rain. Sea breezes can also instigate a persistent line of cloud and rain over a peninsula. Where two sea breezes meet from either side of a peninsula, a long, thin region of convergence forms, pushing air upwards and leading to lines of convection that can persist for many hours (Fig. 13.1A). Lines of showers are sometimes seen in the UK across Devon and Cornwall on a summer's day. In the tropics, the line of convergence caused by colliding sea breezes over central Florida can often instigate deep convection and thunderstorms.

At night, both the land and sea cool down by emission of thermal radiation, with the land cooling down more rapidly. After a time, the temperature gradient across the coastline reverses, with the air over the land becoming cooler than the air over the sea. This instigates a circulation in the opposite direction to the sea breeze, with a gentle wind blowing from the land to the sea, referred to as a **land breeze** (Fig. 13.1B). In the absence of solar heating, this land breeze is typically much weaker than the sea breeze, and rarely strong enough to produce clouds.

The effects of the coastline on weather are not just limited to sea breezes. The contrasts between land and sea can also impact the weather when there is large-scale flow brought about by passing weather systems. Often a coastline can mark the beginning of cloud formation. For example, a capping temperature inversion may be suppressing convection in a mass of air sweeping across the sea. Once the air reaches the warmer land, the extra heating from below can give it the push it needs to overcome this inversion, with the result that clouds can start to form over land (Fig. 13.2A). Conversely, a coastline can cause the cessation of cloud formation; particularly if clouds are precipitating, the passage from sea to land can cut off their moisture supply, leading to their dissipation.

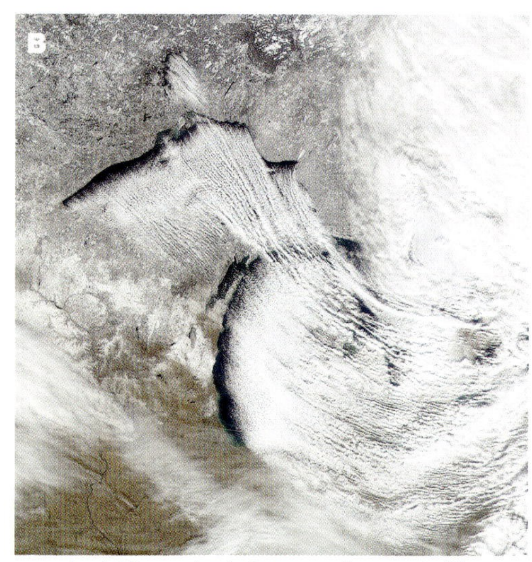

Figure 13.2 (**A**) Cloud streets forming when moist air passes over land. These clouds form over the coast of the British Isles on 24 April 2008. (**B**) Lake effect snow forming over Lake Superior and Lake Michigan on 05 December 2000. (Images: NEODAAS/University of Dundee (A); NASA Earth Observatory (B).)

In winter, land–sea contrasts can lead to a phenomenon called **lake effect snow** (Fig. 13.2B). This is most prevalent over the Great Lakes of North America, although it can occur to a lesser extent elsewhere in the upper mid-latitudes. It happens when very cold winds from over land pass over the water surface. Water surfaces in winter will often be much warmer than the surrounding land, hence the air passing over it becomes heated from below and fed with moisture, destabilising the atmosphere and leading to an outbreak of convection. This convection can give very heavy snowfall to the land on the downwind side of the lake. If conditions are right, convective cells can continuously form over the lake and stream off over the land, bringing several metres of snow to the US states to the south-east of the Great Lakes – Michigan, Ohio, Pennsylvania and New York.

A line of cloud and localised precipitation can sometimes form when the flow of wind is aligned parallel to the coast. This line runs parallel to the coast, and is caused by surface friction between the air and the land. As the

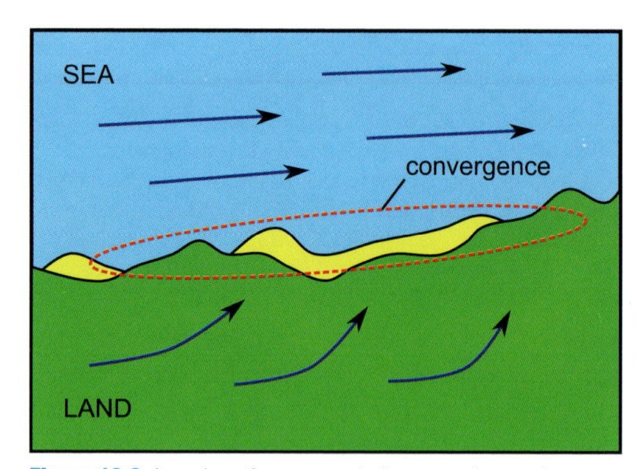

Figure 13.3 Land surfaces exert drag on air passing over them – much more so than the smooth sea. This drag turns the wind to the left, and can cause regions of convergence over the coast.

sea presents a much smoother surface to the atmosphere than the land, the wind tracking over the land is therefore slowed down and deflected to the left (Fig. 13.3). If the land is on the right with respect to the flow, a narrow band of convergence is created, leading to cloud formation. In the converse case, with land to the left, a band of divergence would suppress cloud formation.

13.2 Mountain Weather

A range of mountains presents a solid barrier to any large-scale wind bearing down on them. This leaves the air only one way to go: upwards over the top. The effects of mountains on weather are felt not just in the mountainous region, but also a fair distance to the leeward side. A long range of mountains, such as the Rockies or Himalayas, can have an influence on the entire global circulation. A small cluster of hills and mountains (or **orography**) could easily affect the weather a few hundred kilometres downwind. The flow on the leeward side is often disturbed, giving rise to localised streams of vortices and waves (see Figs 13.4 and 13.5A). Sometimes, clouds can form on the crests of these waves. Downwind of the mountain, gravity waves may stream off the mountain top, with the air flow undulating up and down. If the crests of these waves penetrate the lifting condensation level, the air will saturate, forming lines of wave clouds called **altocumulus lenticularis** (Fig. 13.5B).

The passage of air over hills and mountains can also lead to rapid cloud development. Ascending air can quickly be raised to its lifting condensation level, resulting in cloud formation. These clouds can completely enshroud the tops of the mountains if the lifting condensation level is low enough. The

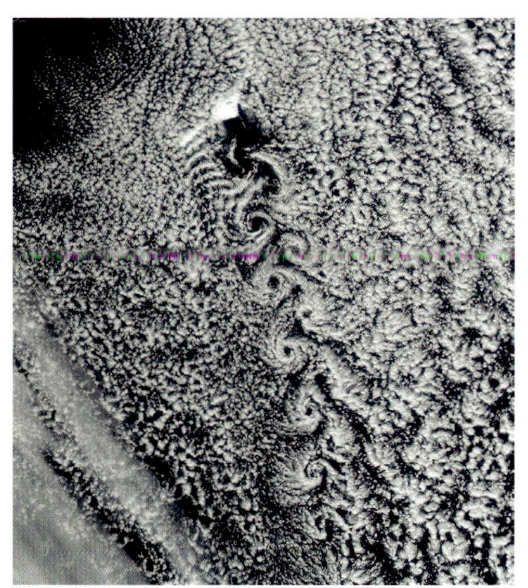

Figure 13.4 The influence of orography on an air stream. Here, vortices have formed in the clouds downstream of Jan Mayen Island in the North Atlantic. (Image: NASA Earth Observatory.)

nature of clouds that form over the mountains is dependent on the large-scale weather situation, although in a situation where a stable layer is acting as an inhibitor to deep convection, the forced ascent of air over a range of hills or mountains can sometimes be enough to release the convection, resulting in very heavy rain over the mountain regions.

If clouds that are already precipitating arrive at a mountain range, the result of the forced ascent is an increase in condensation rate and a marked intensification of the rain. This process is referred to as **orographic enhancement**, and can bring vast amounts of rain to upland areas. A second process can also massively enhance rain over hills and mountains: the **seeder-feeder effect** (Fig. 13.6). If cloud forms over a mountain summit – a so-called

Figure 13.5 (**A**) Waves in the base of a stratus layer over the Alps in July 2011. (**B**) Altocumulus lenticularis over Stockton-on-Tees, UK. Both effects are caused by undulations in the air flow downstream of hills or mountains. (Photos: Claire Delsol (A), John Lawson (B).)

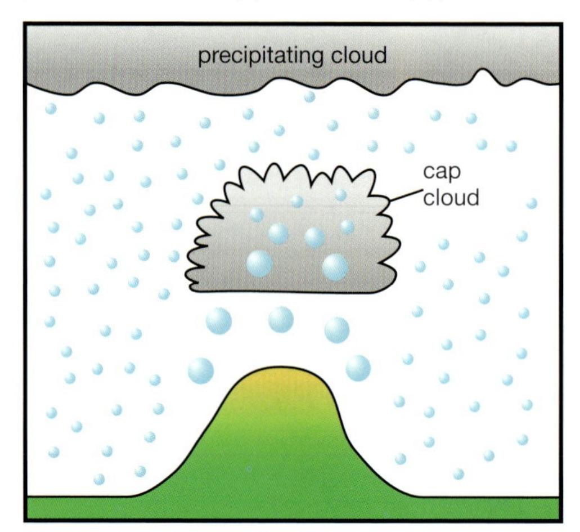

Figure 13.6 Raindrops falling through a cap cloud over the summit of a hill grow as they sweep out the cloud droplets. This is the so-called seeder-feeder effect, and it can massively enhance rainfall over mountainous or hilly regions.

cap cloud – beneath the large-scale precipitating cloud, rain from the large-scale cloud then falls into this cap cloud. As it does, the falling raindrops collect extra water from the cap cloud, resulting in the growth of the falling raindrops. In combination, these two effects can deliver very large amounts of precipitation in a short space of time, with the potential to cause severe localised flooding.

There are two principal types of flooding. The less severe type, **pluvial flooding**, occurs when large amounts of rain fall into the upper reaches of a river's catchment. More water has to flow along the river to clear this excess. If the amount is too great, a river can burst its banks, inundating the local area. A second type of flooding is called **fluvial flooding**. This occurs when the ground becomes so saturated that it can no longer absorb the rain as it falls, with the result that it simply flows off the surface and straight into the rivers, expanding them rapidly. Such flooding is a challenge to forecast, as it has the potential to happen so quickly.

The effects of mountains on weather are, of course, not entirely malign. In some instances, mountains can provide much fairer weather. Saturated air ascending the windward side of a mountain range rises at the saturated adiabatic lapse rate as cloud forms and precipitates. By the time it arrives at the leeward side of the range, it is much drier, with most of its water content having been precipitated out among the mountains. As it descends the leeward side, it warms at the dry adiabatic lapse rate. This means that, by the time it reaches sea level on the leeward side of the range, it can be several degrees warmer than on the windward side (Fig. 13.7). This is referred to as the **Föhn effect** and provides several areas of the world with warm, dry winds. The Chinook wind that blows from the Rockies across North America is an example of a Föhn wind, and its warmth can cause massive increases in temperature within just a short time. Also, as the air is drier, less precipitation tends to fall in the **rain shadow** behind a mountain range.

Local variability in temperature can also occur in mountain regions. For example, in the northern hemisphere, a south-facing slope in a mountain valley will receive much more solar radiation over the course of a day than a north-facing slope. This small difference can lead to substantial differences in local climate, often with one side of a valley being substantially warmer than the other. This can also have an effect on the local snow budget: a sheltered side of a mountain will have much less snow melt than a side exposed to the Sun. Often in spring, the last remaining patches of snow on northern hemisphere mountains are on the upper north-facing slopes.

Localised warming and cooling on mountain slopes can also have local effects on air density. Rapid cooling of mountainsides at night can chill the air just above the surface. This makes the air colder and denser, allowing it to flow strongly downhill, creating a **katabatic wind**. The result of katabatic flow is the pooling of very cold air in the valleys below. In winter, these cold katabatic winds

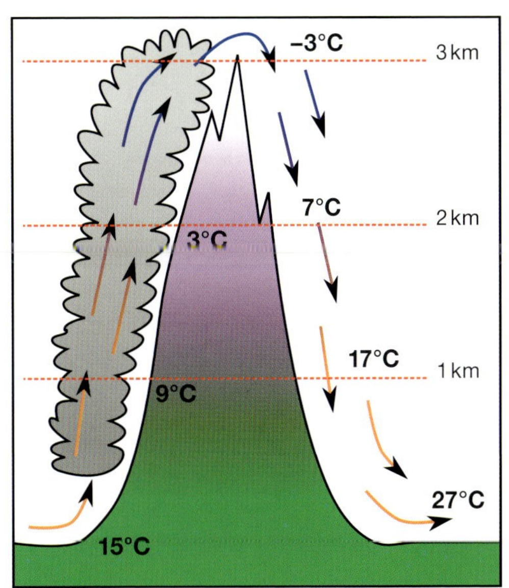

Figure 13.7 If cloud is forming, air ascending the windward side of a mountain range cools at the saturated adiabatic lapse rate (about 6 °C km^{-1}). As it descends down the leeward side, it warms at the dry adiabatic lapse rate, as cloud formation is no longer occurring. By the time it reaches sea level again, it can be several degrees warmer. This flow of warm air off the mountains is called a Föhn wind.

flow persistently down the steep edges of the Antarctic continent, making the coast of Antarctica one of the windiest places on Earth. Conversely, in regions of local warming on a mountainside, the less dense air can start to flow uphill, creating an **anabatic wind.** These tend to be less strong than their katabatic equivalents, although where they meet at the tops of mountains, they can cause local convergence, leading to convection.

13.3 Desert Weather

In contrast to mountain ranges, deserts are identified by a lack of rain. Often, desert regions form in the rain-shadow behind a mountain range. For example, the Atacama Desert in Chile is flanked on either side by mountains, meaning that, whatever direction the wind is coming from, any moisture is rained out before it reaches there. Deserts also occur in the sub-tropics, where low-level divergence inhibits cloud formation and high pressure dominates. There are two main types of desert: **hot deserts** and **cold deserts**. Hot deserts are normally found in the tropics and have rocky, gravelly or sandy surfaces and are marked by large swings in temperature between day and night on account of the lack of cloud. The dryness of the surface means that solar radiation can be used to heat the surface all day, as opposed to evaporating moisture. Clear skies mean that this heating is uninterrupted from sunrise to sunset, when rapid thermal cooling leads to substantial drops in temperature. In contrast, cold deserts are found at higher latitudes. Some cold deserts also have rocky surfaces, although some have icy surfaces – the largest desert in the world is the interior of Antarctica.

Despite the lack of rainfall, the strong temperature contrasts between day and night can still cause strong winds in desert areas. These tend to happen around the desert margins, where there is some moisture and the potential for clouds and rain to form. A mature convective system travelling over a dusty surface can kick up a great deal of dust within vigorous circulation near the ground. Downdraughts from the cloud can blast dust off the surface, creating a wall of dust along the edge of the gust front. This wall is referred to as a **haboob**, seen by an observer as an ominous wall of dust approaching at speed. Dust storms can also occur in areas ravaged by drought. If there has been no rain, the Earth's surface will be very dry and dusty; hence any strong wind has the ability to lift the dust into the atmosphere. This can make drought conditions even more severe; drought itself can damage crops, livestock and livelihood, but breathing dust-laden air is not healthy and can lead to all manner of respiratory problems. During the 1930s, drought in the southern USA led to the Dust Bowl – a series of dust storms that swept plumes of dust across vast areas, depositing it as far away as New York.

The intense heat of the desert day can also give rise to small vortices that spin off the desert surface, carrying dust upwards. These are referred to as **dust devils**. They are rarely strong enough to do any major damage – certainly far weaker than the tornadoes discussed in the previous chapter.

Dusty air lifted to high altitudes from the interior of a desert region can also be transported over many hundreds of kilometres and deposited in some distant area. For example, dust lifted from the Sahara can be transported westwards off the African coast (Fig. 13.8). If it becomes wrapped up in mid-latitude depressions, the dust can be carried across Europe and dropped in precipitation. Dust can also be transported out of northern Africa in low-level winds, although not to the same extent. If there is low pressure over the Mediterranean, hot, dry, dusty air can be drawn up over Europe from North Africa. Depending on the location of the wind, it can be known as a **sirocco** (from the Algerian and Tunisian coasts, influencing Italy and southern France), a **khamsin** (from Egypt across the eastern Mediterranean) or a **leste** or **leveche** (from Morocco to the Iberian Peninsula).

Figure 13.8 Dust from the Sahara being transported out over the Atlantic Ocean on 8 March 2012. Note the ripples in the dust in the lee of the Canary Islands. (Image: NASA Earth Observatory.)

13.4 Urban Weather

The scale of a town or city compared to the naturally forged features of coastlines, mountains or deserts is small. Despite this, the presence of an urban area can have a sizeable impact on the weather conditions of the surrounding area. Tall buildings have a marked effect on wind – indeed, in a similar way to a mountain range, although on a much smaller scale. Wind is deflected and turned as it passes through a built-up area. A street flanked by tall buildings can act as a channel that redirects wind along it – for this reason, such streets are often referred to as **street canyons**. Wind travelling along a street canyon may be moving in a very different direction to the background wind direction.

The nature of the flow of wind through these street canyons is a topic undergoing extensive research, as it is an important part of understanding the dispersion of urban pollution. Busy streets can contain traffic that emits a range of pollutants into the atmosphere. However, while this could easily be

Figure 13.9 Smog over São Paulo, Brazil, trapped under a temperature inversion. (Photo: Wagner Nogueira Neto/ Ana Carolina Lacorte de Assis.)

dispersed by the wind in open countryside, the flow within a street canyon can trap pollutants, perhaps causing them to pool at street level, which can present a health hazard for inhabitants of the urban area. Pollutants over an urban area can also lead to a dirty layer of cloud called **smog**. An excess of pollutant particles and aerosols over a city can enhance the possibility of cloud formation, but any cloud that forms in this heavily polluted air is likely to be dirty. When there is a strong temperature inversion above a city, this smog can gather below it and persist for hours (Fig. 13.9).

The air temperature is often a few degrees higher in urban areas than in the surrounding countryside. This is often the case both during the day and at night, and is referred to as the **urban heat island effect** (Fig. 13.10). There are a number of factors that contribute to this extra warmth. Firstly, cities tend to be covered by vast areas of tarmac, which is a much darker surface with a lower albedo than the surrounding fields. This means that the urban area both absorbs and emits more radiation. Secondly, the surface area of the city centre is greater: in the countryside, thermal radiation is only transferred between the ground and the atmosphere above; in the city, building sides present extra area through which thermal radiation can be transferred. Thirdly, cities are usually designed to channel water away, in contrast to the soil surface of the countryside,

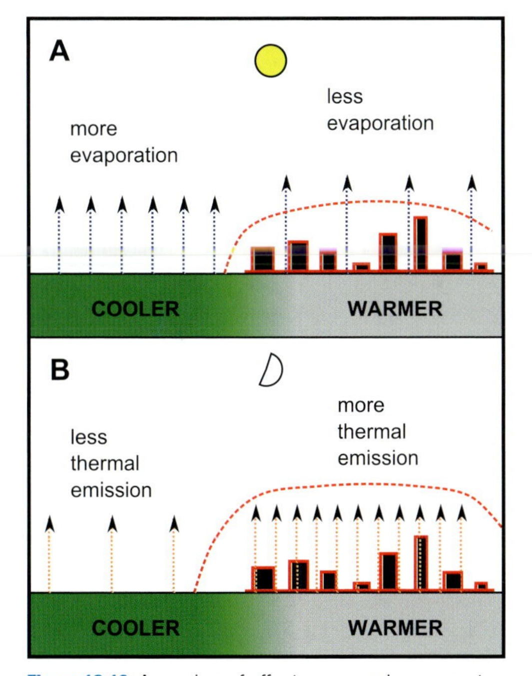

Figure 13.10 A number of effects cause urban areas to be slightly warmer than the surrounding countryside (the so-called urban heat island effect). The principal effect during the day is differences in evaporation rates; at night, the increased thermal emission of the urban landscape is an important factor.

which can store moisture. Put together, this implies that more of the incoming solar radiation is absorbed by the urban surface and less of the absorbed energy has to be used to evaporate moisture. This extra energy can be used to heat the air, resulting in warmer conditions than the countryside. In addition, the city also produces its own heat – for example, heating systems in buildings and car engines both produce extra thermal energy.

The urban heat island, however, is most evident at night. During the course of the preceding day, a great deal of energy can be absorbed by the city surfaces. At night, both urban and rural areas cool down as thermal radiation is emitted. Also, a fourth effect – the design of buildings – comes into play. Most buildings in the mid-latitudes are designed to store heat so that their inhabitants can stay warm. This means that the buildings themselves can radiate thermal energy, further heating the city. These effects can allow the night-time temperatures in the city to remain several degrees warmer than in the surrounding countryside.

14 Forecasting the Weather

Since time immemorial, mankind has tried to predict the weather. We have already seen in Chapter 2 the many different, subjective ways that forecasts have been made in the past, making use of natural indicators. Of course, we have come a long way since. The forecasting techniques of Robert Fitzroy and Lewis Fry Richardson helped pave the way for modern forecasting, and since then the meteorological world has been striving to improve its ability to forecast the weather. The development of modern computational techniques has allowed forecasters today to give a much more reliable prediction of the coming weather than back in Fitzroy's day, when forecasting involved poring over weather maps and looking for patterns.

The process of generating a weather forecast brings together every area of meteorology. Intensive observations of the Earth–atmosphere system are needed; the background physics describing how the atmosphere behaves needs to be applied; and the processes by which weather patterns evolve need to be understood. Every day, weather agencies around the world feed weather observations into computer models, which apply the background physics to the observations and project the state of the atmosphere forward in time. These weather models form the core of modern weather forecasting, and it is their output that is used by a forecaster to issue predictions of the forthcoming weather.

14.1 Numerical Weather Prediction

The basic principle of using physical equations to project a grid of observational weather data forward in time is referred to as **numerical weather prediction**. Despite the many advances since, numerical weather prediction still follows the basic principles of Richardson's first forecast. His equations described simple physical laws of thermodynamics and fluid mechanics, formulated a few decades earlier by Vilhelm Bjerknes. Richardson took a snapshot of weather data from across Europe and divided the area into a number of gridboxes, bordered on the sides by lines of longitude and latitude. Based on the reported values in each gridbox, he manually calculated the change in pressure over a six-hour period for two atmospheric columns from the surrounding values.

Within a decade of the first successful numerical forecast, run on ENIAC in 1950, national weather agencies around the world were obtaining supercomputers with the aim of developing and running forecast models on them. The Swedish Air Force was running forecasts as early as 1954, with the USA following suit a year later. The UK Met Office purchased its first computer in 1959 and began forecasting operationally in 1965. As the computers of the time were considerably slower than modern ones, the forecasts they produced were simple and predicted only a few meteorological variables, and covered the globe with a small number of large gridboxes. Such simple forecasts left a great deal of interpretation to be done by a human forecaster for the forecast to be useful to the public.

Since the days of ENIAC, the increase in computer speed has been phenomenal. Running flat out, ENIAC could perform about 500 floating-point operations per second. To put this into perspective, modern supercomputers are reaching processing rates of 10^{15} floating-point operations per second (Fig. 14.1). With this extra processing capability, it has been possible to incorporate increasing amounts of detail in our weather models. We can now include many more meteorological variables in our models over a much increased number of gridboxes. In addition, a great deal of research has been put into ways of improving how different aspects of the weather are treated within the model, allowing the development of more realistic representations. On top of this, a rapidly growing observation network has provided a greater detail of data to verify these representations. The result is an increase in success rates of weather forecasts – nowadays, five-day forecasts have about the same success rate as three-day forecasts did 20 years ago.

Figure 14.1 This IBM supercomputer is currently in use at the UK Met Office. Like ENIAC, modern supercomputers still fill a whole room, but can perform calculations at a much faster rate. (Photo: Met Office.)

However, even with our most modern supercomputers, forecasts still sometimes go wrong. Weather forecasting is a huge, complex and very difficult mathematical challenge. Reaching the state where we have a system that could accurately predict the global weather conditions with 100% reliability is impossible: to stand a chance of creating a perfect forecast, we would need to know the exact state of every molecule of the entire atmosphere at a given time, and then build a model that could predict exactly how all these molecules interact. Clearly, even with modern supercomputers, none of this is practical. The data storage and processing time required to model all 2×10^{44} molecules of the atmosphere would be unfathomable. In short, we have no choice but to accept that we cannot have a perfect forecast system. Instead, we must make a number of assumptions when building our weather model, and ensure that these assumptions are neither so complex that they take too long for the model to calculate, nor so simple that the modelled weather behaves in an unrealistic way.

14.2 The Initial Conditions

Before we can begin to run a weather forecast on a supercomputer, we need information about the state of the atmosphere at a given time – a global snapshot of weather observations. This snapshot is referred to as an **analysis** and forms the initial conditions for the forecast model. To build this snapshot, we use our global weather observation network. We saw in Chapters 3 and 4 how much information is available to us from surface observations, vertical soundings by radiosonde, radar observations of rainfall, and satellite soundings of the atmosphere. The vast majority of the observational information that forms our analysis is actually derived from satellites, which can remotely monitor not just cloud patterns, but also variables such as pressure, temperature, humidity and wind speeds.

These observations must be mapped onto a grid before they can be used. The entire surface of the globe is divided into gridboxes, normally bordered by lines of longitude and latitude (Fig. 14.2). The column of atmosphere above each of these gridboxes is divided into vertical layers, extending from the Earth's surface all the way up to the upper atmosphere. It is usual to pack most of the layers into the troposphere, where most of the weather occurs, with the narrowest layers nearest to the surface. Layer boundaries are defined to follow the terrain of the surface, so that no gridbox extends under the ground (Fig. 14.3).

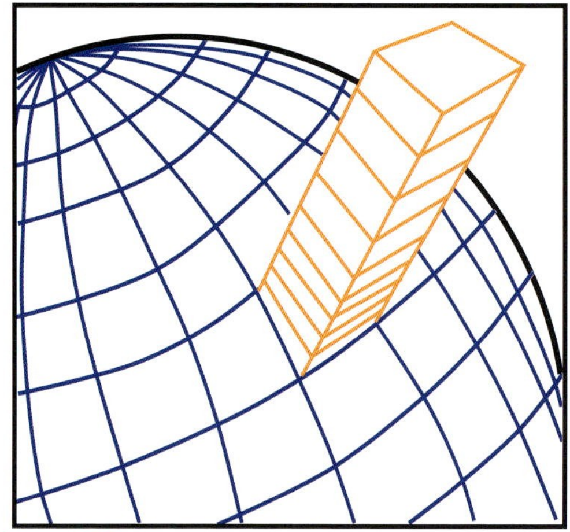

Figure 14.2 Most models divide the world into gridboxes, bounded by lines of longitude and latitude. The atmosphere in each gridbox is then further divided into layers, with more layers in the lowest reaches of the troposphere.

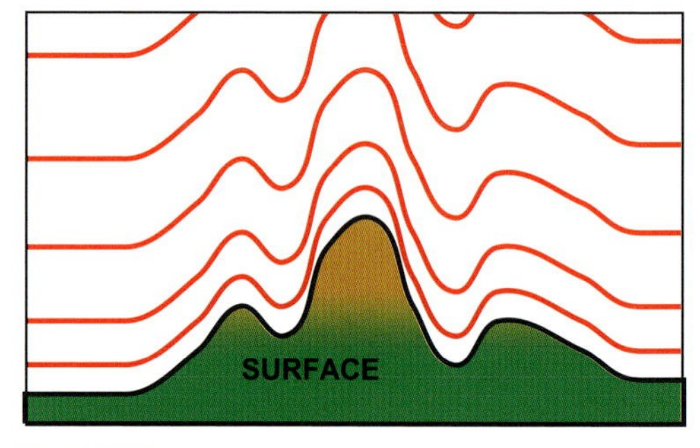

Figure 14.3 Model levels track over orography so that no model gridboxes extend below the ground.

The initial conditions require information on the state of the atmosphere in every layer of every gridbox. However, we do not necessarily have observations of the atmosphere in such detail; we have already seen how unevenly distributed the global observation network is. This presents us with a problem, as we cannot run our forecast model with gaps in the data. One possible solution is **interpolation**, which involves filling in these gaps by assuming that the atmospheric state varies smoothly between neighbouring observations. However, this method has the potential to miss important features, particularly over the oceans, where observations are sparse and many types of weather systems begin their lives.

To fill the gaps, we make use of output from a previous run of the forecast model. For example, say we have a global snapshot of observations at 12:00 UTC. At this time, we also have a six-hour forecast of the conditions at 12:00 UTC that was run from the analysis produced at 06:00 UTC, which contains all the required information about the state of the atmosphere at every point. The challenge is then to merge the six-hour forecast with the observations. They are unlikely to agree exactly – for a start, even after only six hours, there may be minor disagreements between the forecast fields and the actual state of the atmosphere. We cannot simply copy our observations directly into the forecast field. For the forecast to run, the fields need to be physically realistic. For example, copying a surface pressure observation of 1,025 hPa into a single gridbox that is surrounded by pressures of 1,020 hPa in the forecast field would result in an unrealistically sharp pressure gradient. Running a forecast model with such sharp gradients can lead to the model generating atmospheric waves in the model that can lead to large, unrealistic swings in pressure and temperature. Such waves are probably what caused Richardson's first forecast to predict such a large pressure change.

Combining the two fields in a way that takes all of the observations into account while remaining physically feasible is a challenging mathematical task, referred to as **data assimilation**. This is a complex process that takes the forecast field and nudges it back towards the observed values. Care must be taken during this nudging process, as it is very possible that spurious measurements may have crept into our observational field, perhaps caused by a malfunctioning automatic weather station or a defective radiosonde. It is crucial that such data is removed from the process. Computerised data assimilation algorithms seek out these anomalous values by looking for any unrealistic variations in the fields. Any potential errors are flagged up, and can then be checked by a forecaster. The combination of the previous forecast fields with the latest observation fields forms our analysis – a complete global snapshot of the weather at all gridpoints and all layers. This analysis can be used as the initial conditions for the next forecast. A typical forecast cycle is depicted in Figure 14.4

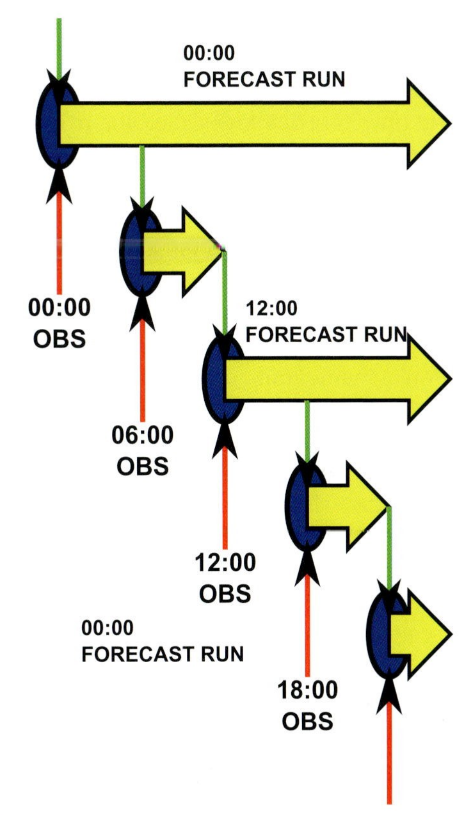

Figure 14.4 A typical forecast cycle over 24 hours. Every six hours, an analysis from the previous forecast run is assimilated with the six-hourly observations to initiate the next forecast. Each 12 hours, a full forecast run is performed; at 06:00 and 18:00, short forecast runs are performed to provide an analysis for the next forecast run.

14.3 Running the Model

A modern weather forecast model is a massive computer program made up of millions of lines of code spread over hundreds of subroutines. Performing even a short run of this forecast model from a set of initial conditions requires a copious number of calculations to be performed. Operational weather forecasts can therefore only be run on the world's fastest supercomputers.

The heart of a forecast model is a set of basic physical equations called the **primitive equations**, which are still based on the set of equations first derived in the early twentieth century by Vilhelm Bjerknes. They are a set of partial differential equations, describing how pressure, temperature, wind and water content vary with time and space, using simple rules of physics, including the conservation of momentum, continuity (the fact that air cannot be created or destroyed) and simple thermodynamics. But to be useful in our forecast model, the primitive equations must be modified slightly. In our model, we deal only in spatial grids of finite horizontal and vertical spacing, and we advance our model in discrete time steps, so we need to convert the equations into a form that deals with finite steps in space and time. This process is called **discretisation** and leads to a set of equations that can be used to project the state of the atmosphere forwards in time.

With the primitive equations alone, a forecast model is only capable of predicting a few meteorological variables. For example, while they can predict temperature, they cannot directly model the processes that lead to cloud formation. Modern forecast models therefore contain many additional sections that link the primitive equations to other aspects of the weather. A typical model will contain a radiation scheme to calculate how both solar and terrestrial radiation are redistributed throughout the atmosphere, and a cloud scheme that uses information on the amount of water vapour in a gridbox to determine where clouds might form. Many also contain more detailed schemes to represent the lowest layers of the troposphere (the **boundary layer**), which is an important factor in the complicated transfer of heat and moisture between the surface and the free atmosphere above.

Some of these processes occur on scales much smaller than the size of a gridbox. Such processes therefore cannot be directly represented by the model, and must be represented statistically, or **parameterised**. For example, in a weather model of gridbox size 25 km, it is not possible to depict exactly where individual cumulus clouds of size 1 km might develop. However, by analysing the distribution of water vapour in each layer of the atmosphere and the stability of the atmosphere, it is possible to predict whether or not cumulus clouds will develop, and to what extent. Similarly, we can analyse the number of layers these clouds extend through and, if they become large enough, predict the presence of rainfall.

Of course, the more detail we put into our model (both in terms of the number of processes represented and the number of gridboxes), the longer it will take to run our forecast. Using smaller gridboxes allows us to better represent local features of the weather, such as clouds and rainfall, but at the cost of increasing the number of calculations we must perform in a model time step. But clearly a weather forecast model that takes seven hours to generate a six-hour forecast is never going to be of any use. We need to set up our model so that it can produce forecasts for potentially several days ahead over the course of maybe just a few hours. To achieve this, operational forecast models use a global longitude–latitude grid with a gridbox size of about 25 km in the mid-latitudes and 60 to 100 vertical layers.

We must also choose our time step carefully. If the time step is too short, the forecast will again take too long to run; if the time step is too long, the forecast can fail by becoming **unstable**. The equations used in the model only allow an individual gridbox to communicate with its direct neighbours. Over short periods of time, this is reasonable. However, over longer periods of time, a gridbox may become increasingly influenced by gridboxes beyond its neighbours – something that the primitive equations would miss out on, leading to the model potentially producing unrealistic results. Time steps in operational forecast models tend to be in the range 1 to 10 minutes, depending on the horizontal spacing of the grid.

14.4 Global and Regional Models

National weather services typically run operational forecasts of the next seven days or so every 12 hours (at 00:00 and 12:00 UTC) with each forecast run taking a few hours. Shorter forecast runs are usually performed at 06:00 and 18:00 UTC to provide forecast analyses that can be combined with the observations to give the full analysis for the next forecast run. However, the output of the 25 km **global model** alone only provides a forecaster with an overview of what is likely to happen. It can give reasonable forecasts of winds, pressure and temperature and can probably indicate areas where cloud and rain may develop. However, for more regional forecasting, more precise information on the coming weather is desirable. In other words, a smaller gridbox size is ideally required. But reducing the size of gridboxes across the entire global model would massively increase the time taken for the forecast to be run.

The alternative is to run a **regional model**. There are two main differences between global and regional models. Firstly, global models cover the entire Earth, while regional models only cover a specific part of the Earth, referred to as the model **domain**. Secondly, as regional models cover a smaller area, they can have smaller gridboxes, enabling them to better resolve local weather features than the global model. Aside from those differences, the physical equations used to run both types of model are exactly the same.

A challenge to running a regional model is how to deal with the **boundary conditions**. The global model has the advantage that, as it is modelling the whole atmosphere, there are no boundaries – a weather system tracking to the east could theoretically keep on travelling from gridbox to gridbox, circumnavigating the Earth. As the regional model only covers a small area of the Earth's surface, it must have boundaries on all sides. A weather system travelling eastward could easily travel out of the domain on the eastern side. But what happens when a system enters the domain from the western side? To run, the regional model needs a constant supply of information about any weather systems entering the domain on any side. This information can be taken from the global model during its run. During an operational forecast run, global and regional models usually run at the same time, with the global model constantly passing information down to the regional model to fix its boundaries. Most forecast agencies run a suite of models at different resolution and domain size, each one providing boundary conditions to the next (Fig. 14.5). For example, at the UK Met Office, a global 25 km model feeds a regional model that covers Europe and the North Atlantic at

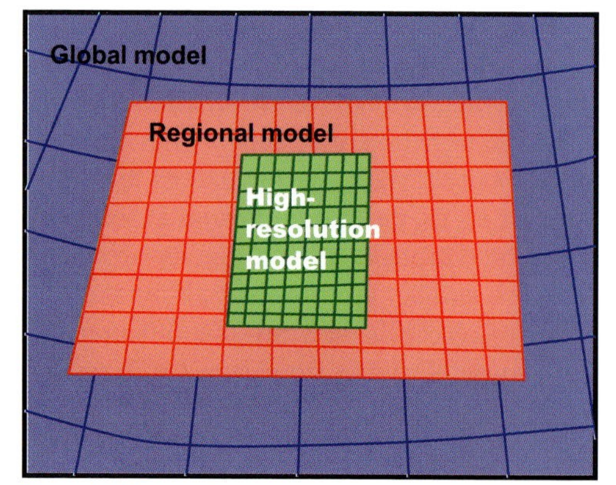

Figure 14.5 Nested grids allow high-resolution versions of forecast models to be run over areas of interest. Here, the regional model marked in red sets its boundary conditions from the global model marked in blue. In turn, the high-resolution model in green takes its boundary conditions from the regional model.

a resolution of 12 km. In turn, this then feeds down into a second regional model, covering the UK and Ireland at a resolution of 4 km and a third one at a resolution of 1.5 km.

14.5 Ensemble Forecasting

So, weather models provide forecasters with a single prediction of how the weather may evolve over the next week or so. But we know that forecasts vary in terms of how well they predict reality, and the output may be right or wrong. This is little consolation if a forecast of dry weather is given for a period when heavy rain falls. In many instances, it is far more useful to give a forecast of the chances of a certain weather event occurring. But how do we obtain probabilities from a weather model?

The answer lies in the principle of **ensemble forecasting**, which is performed by many weather agencies around the world. A single weather forecast is run using the operational model. Then, in addition, a number of

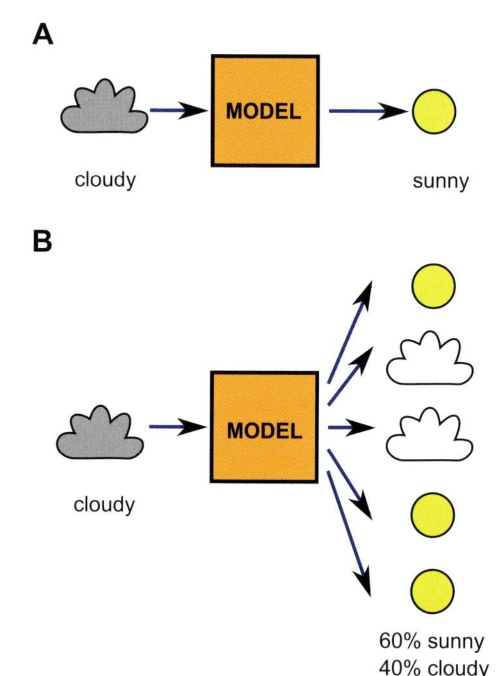

Figure 14.6 A simple schematic to compare a deterministic forecast (**A**) with a probabilistic forecast (**B**). Deterministic forecasts give a single forecast for a single set of initial conditions. In this case, from cloudy initial conditions, the forecast predicts sunshine. Probabilistic forecasts (**B**) use an ensemble, which gives many possible forecasts for a single set of initial conditions – here, the cloudy initial conditions are slightly perturbed within the model to give multiple forecasts that predict a 60% chance of sunshine and a 40% chance of cloud.

other forecasts are run with slightly different initial conditions. We know that there is uncertainty in the weather observations used in the model, and that this uncertainty could lead to large differences in the state of the predicted weather even after just a few days. Hence, if we randomly perturb our initial conditions to account for this uncertainty, we can get a number of possible alternative outcomes for our operational forecast (Fig. 14.6).

This approach allows us to determine the probability of certain weather conditions occurring, and is often called **probabilistic forecasting** (in contrast to **deterministic forecasting**, where only a single forecast run is performed). From an ensemble of 50 forecast runs where 40 predict rain while 10 predict dry conditions, a forecaster can infer that there is a higher chance of it raining than not. However, if he only had a single deterministic forecast of the same situation, he has far less information to go on – indeed, if the single run predicted dry conditions, it is quite likely his forecast of dry conditions will be wrong.

Typical ensemble sizes vary from weather service to weather service. For example, the ECMWF runs a 51-member ensemble; the Met Office runs a 24-member ensemble. As these forecasts are being run in addition to the operational forecast, they are usually run at lower resolution (larger gridboxes). As there is a great deal of information in the output of an ensemble forecast, computer techniques are often used to group together forecasts that are predicting similar conditions. This process is called **clustering**.

15 The Forecaster's Challenge

A supercomputer usually takes a few hours to run the forecast model. This may include a global model coupled to a suite of regional models; it may include a number of low-resolution ensemble forecasts to give an idea of the probability of different types of weather occurring. The result of these model runs is a large amount of output data describing the predicted state of the weather at many time intervals into the future.

It may seem that the job is now done. However, the computer output is only the first step of the forecast process. To convert the raw model output to a meaningful weather forecast that could be interpreted by the public requires a great deal of interpretation – the sort that could only be provided by the judgement of a human forecaster with experience. While weather observation is becoming increasingly automated, it seems unlikely that forecasting will follow the same trend. Even with our modern advances in supercomputer technology, human common sense will still be required to whittle out the erroneous data and issue forecast corrections when the weather forecast produced by the computer model starts to disagree with the actual weather.

15.1 Making a Weather Forecast

Making sense of the output data from the forecast model is certainly not a task for the faint-hearted. Yet, either twice or four times a day, forecasters around the world are challenged with converting model output into meaningful forecasts. During a forecast cycle, a forecaster may be expected to provide forecasts on many scientific levels: for TV forecasts, for example, the information needs to be simple and clear for the public to understand; for military purposes, such as flight briefings for pilots, forecasts must be more formal and scientific; for a newspaper, the forecast may have to be condensed down into ten words. And once the forecast cycle is complete, the whole process begins again as the next round of forecasts is calculated.

Output from the run of a weather forecast model is normally displayed in the form of weather maps (Fig. 15.1), created automatically by computer and called **prognostic charts**. These charts are plotted for every six hours through the forecast period and show the forecaster how different aspects of the weather (such as pressure and temperature) are predicted to evolve. **Meteograms** (Fig. 15.2) are also a useful way of displaying the model output: these show the trends in various quantities over time, and can give an idea of how the model is predicting the weather to change for an individual city. If an ensemble forecast is being run, meteograms can give an indication of the range of uncertainty in the forecast quantities.

The first step of the conversion process from model output into weather forecast is the creation of the weather **guidance**. This is a summary of how the synoptic situation is expected to change over the forecast period,

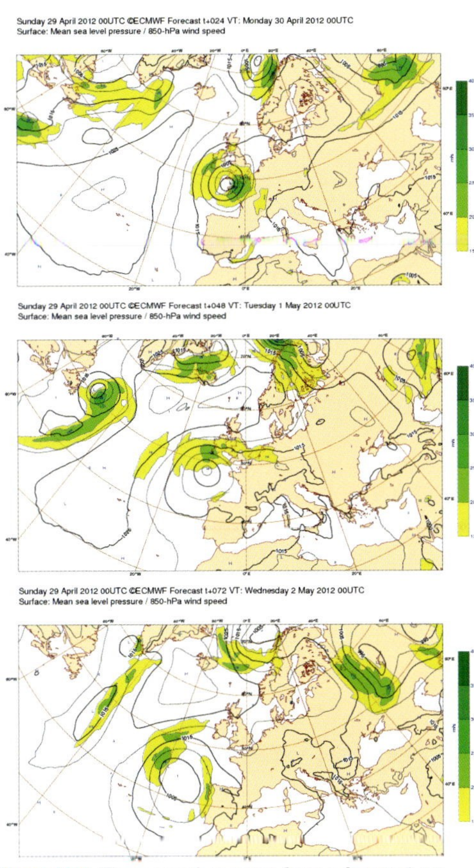

Figure 15.1 Three weather maps from a forecast run from 00:00 on 29 April 2012. (Image: ECMWF.)

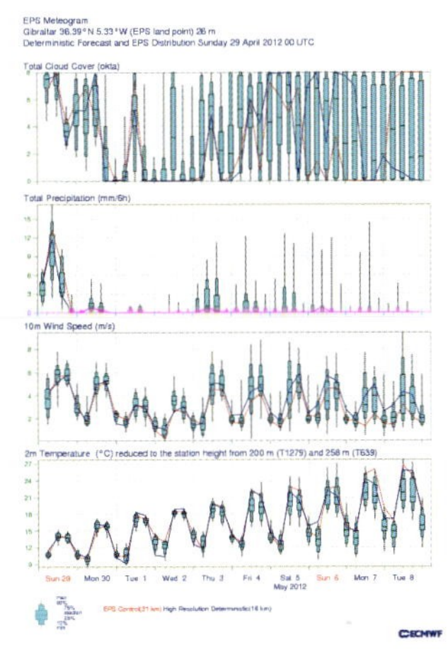

Figure 15.2 Meteogram showing ten-day forecasts of cloud cover, precipitation, wind speed and temperature for Gibraltar from 29 April 2012 (the same forecast run as the maps in Figure 15.1). This shows the output of an ensemble forecast: the width of the bars indicate the likely range of each quantity every six hours. (Image: ECMWF.)

and is compiled by the most senior, experienced forecasters. Often several senior forecasters gather to discuss the guidance, and sometimes even collaborate with forecasters from other weather agencies. This guidance is then passed on to the rest of the forecasters on duty and used as the basis for the weather forecast.

As well as the forecast output, forecasters have a range of other information available to them – weather agencies are constantly being fed the latest weather observations, satellite and radar imagery (Fig. 15.3). This allows the forecaster to constantly monitor the actual weather conditions and compare them with the model output. As a forecast run usually finishes a few hours after its initiation time, there will already be a couple of sets of hourly synoptic observations that can be compared with the model output. If there is good agreement between observations and model in these first few hours, the forecaster can be more confident in what the model is predicting. If, however, after a few hours there are already significant differences, the forecaster may hold the predictions of the model in less regard.

By international agreement, forecasters at national weather services also have access to model output from several other countries. If the model at their own agency is doing a

Figure 15.3 Modern forecasters are surrounded by monitors that provide them with access to all the latest observations and forecast data. (Photo: Met Office.)

poor job at predicting the weather on a given day, they can have a look at the predictions from these other models. While each of these models is based on the same physical principles, the way they deal with and represent certain aspects of the weather can be very different – indeed, a pair of models can sometimes give two different forecasts from the same initial conditions. With time, forecasters will gain a feeling for which models are more or less reliable in a given weather pattern.

Access to satellite and radar imagery enables forecasters to perform simple short-range forecasting known as **nowcasting**. This involves looking at the current trends in weather conditions and using them to predict the weather ahead – usually just over the next few hours. It also allows the forecasters to remain completely up to date with the current weather conditions and to check that the forecast model output is still in line with the actual weather throughout the whole forecast cycle. Indeed, if the prediction of the next few hours becomes very different to the forecast issued at the previous 6-hour or 12-hour interval, it may be necessary to change the guidance and issue an amendment to the forecast.

The interpretation of local weather conditions is a part of forecasting for which human judgement is critical. Whether the forecaster has model data from a 25 km global model or a regional model of resolution 4 km, the model is never going to be able to give any indications of what is happening on a smaller scale within each model gridbox. For a larger gridbox, this is particularly true: a 25 km box may contain a combination of rural and urban surfaces; coastal boxes contain both land and sea; some may contain hills or mountains too small to be individually represented. A gridbox surface temperature prediction of 10°C may be up to 12°C over the urban area on account of the urban heat island effect (see Chapter 13). A forecaster may note, for example, that the maximum temperature in the urban area is systematically 2°C higher than the surroundings in certain weather conditions, and will therefore need to modify the forecast accordingly.

15.2 Forecasting Hazardous Weather

Every day throws up a brand new set of weather conditions for weather agencies to forecast. In most instances, the forecaster's challenge may only extend to predictions of whether it will rain or not, or whether it will be sunny or cloudy. In society, such forecasts are an important part of daily life, leading us to make simple decisions on what to wear or whether to take an umbrella to work. Sometimes, however, the weather can become extreme. When this happens the forecaster's job becomes even more important – hazardous weather conditions could lead to lives being endangered. It is therefore vital to issue adequate, clear warnings to the public to take action if necessary, if there is a threat of severe weather.

Most national forecast agencies have systems of weather warnings in place. The types of warnings they produce vary depending on the types of hazardous weather that are likely to strike in that location. For example, the UK Met Office warnings include heavy rain, snow, fog, strong wind and cold weather; warnings that can be issued by the US National Weather Service include severe winter weather, thunderstorms, tornadoes and wildfires.

During periods of extreme weather the principle of nowcasting becomes all the more important. Any deviations between forecasts and observations need to be flagged up as soon as possible – a small difference could mean a storm striking land in a different area, or a band of heavy snow developing more rapidly. In severe weather, perhaps the greatest challenge to the forecaster is when to issue the weather warning. A hazardous weather event is far more likely to happen if it appears in a one-day forecast than in a three-day forecast. However, if the forecaster waits until the event is virtually certain to happen before giving a warning, the warning may only be issued a short time before the severe weather arrives, not giving the public enough time to react. Conversely, if he issues the forecast too early, he runs the risk of having to change or remove the warning if the severe weather never actually happens. To counter this, weather agencies tend to have different levels of weather warning, depending on the certainty of severe weather occurring. These range from weather watches, advising the public to be aware of the risk of hazardous weather, all the way up to full-blown weather warnings, recommending immediate action.

Many types of natural hazard actually fall beyond the remit of conventional weather models. For example, while a weather model may be able to predict large amounts of heavy rain over an area, it cannot predict whether this rain could lead to flooding (Fig. 15.4). Similarly, it may indicate wind speed and direction at sea, but cannot predict how the wind might affect the motions of waves. In such instances, additional computer models need to be used. Modern forecast models are designed to be as versatile and adaptable as possible, so that their output can be easily fed into other models of different parts of the Earth system. For example, feeding the model output into a hydrological model, which models how rainfall travels through a river catchment, can give advanced warnings of flooding. Similarly, a model of the ocean surface can provide warnings of storm surges and rough seas – important information for low-lying coastal areas and offshore oil rigs. These are just a few of many types of model that can be used

Figure 15.4 Flood forecasts are a vital part of forecasting. This image shows flooding of the Fitzroy River, inundating Rockhampton, Queensland, Australia, on 09 January 2011. (Image: NASA Earth Observatory/USGS Earth Observing-1.)

outside the forecast model to issue weather warnings.

15.3 Users of Forecasts

Whether a forecast of severe weather or not, forecasts of the weather are important to all of society – and not just the general public. The public rarely obtain weather forecasts from the forecaster themselves: most of the information we receive about the future weather is via the media, the TV forecast being perhaps the most popular. Some TV companies employ their own forecasters to interpret the forecasts from national weather services; some employ forecasters from the weather services directly. Forecasts are also heard on the radio, and usually appear in some form in our daily newspapers. Most TV and radio stations and newspapers also publish their forecasts online, as indeed do most national weather services.

These days, a wide range of companies and businesses need forecasts of the coming weather for a range of different purposes (Fig. 15.5), and it is up to the forecasters to deliver these forecasts as and when they are required. For a start, the energy providers need to know about the coming weather so that they have an idea of whether energy consumption is likely to increase or decrease over the next few days. For example, a period of cold weather will increase energy demand as the public turns its central heating on, while a period of warmer weather is likely to reduce demand. With the recent growth in renewable energy, they also require forecasts to determine how much energy they will be producing from wind farms. The construction industry can be affected by the weather: an unfinished building will need protection from rain, while wind could make working on a tall building dangerous. Tall cranes can be hazardous to operate in windy or stormy weather. Hot weather may also make manual work more strenuous.

Retailers also need an indication of the coming weather, as weather conditions are likely to change the buying habits of customers. For example, hot weather in summer may inspire customers to buy more summer clothes and beachwear, while a cold snap in winter may lead to increased purchases of scarves and woolly hats. A clothing retailer therefore needs to know the weather a few days ahead to make sure they have ordered in enough stock. Similarly, food retailers and supermarkets can tailor their displays to stock more of the types of food that would suit the weather conditions – for example, hot summer days may lead to more sales of burgers and sausages as the public get their barbecues out.

Agriculture is another area of industry that is heavily affected by the weather and needs to be aware of both the current weather and what is on the way. Forecasts of long periods of dry weather mean that crops may need irrigation. At certain times of year, forecasts become even more important: at harvest time, wet weather can be detrimental, as boggy fields could lead to farming equipment becoming stuck and cut crops becoming soggy. Agriculture also makes use of **seasonal forecasts** – farmers need to know if the following few months are likely to be wetter, drier, warmer or colder than average.

With such a high demand for forecasts, weather forecasting has become a lucrative business in some countries. Traditionally, the national weather services around the world were the centre of weather forecasting. Nowadays, however, many independent weather forecasting companies are popping

Figure 15.5 Many businesses, industries and organisations are strongly affected by the weather – for example, (**A**) renewable energy suppliers, (**B**) shipping companies, (**C**) agriculture and (**D**) motor racing teams among many others. Such organisations are therefore very dependent on weather forecasts. (Photos: Jon Shonk (A, D), Jane Shonk (B), Peter Shonk/Alec Murray (C).)

up and selling their forecasts too – generally with the aim of outperforming the national weather services. The forecasting techniques of these companies vary: some employ a panel of forecasters and set them to work on forecast data from national weather services; other companies use secret statistical methods to provide both short-range and seasonal forecasts. As with all forecasting, no agency is capable of providing exact forecasts all of the time, and the quality of the forecasts provided by these companies can be variable.

15.4 When Forecasts Go Wrong

Despite all the data available to the modern weather forecaster, the most powerful weapon in the forecaster's arsenal remains experience. The more forecasts forecasters make, and the more weather situations they see, the more idea they will have as to how the weather is likely to evolve on a given day. Also, the more experience they have, the more they can build up local knowledge of how weather on the large scale can affect conditions on local scales much smaller than the size of a model gridbox. However, occasional wrong forecasts are inevitable. Through the previous chapter we described a number of assumptions that are made when building a weather model. While these assumptions are becoming increasingly realistic as our supercomputers become ever faster, they are still incapable of capturing every circulation and process in the atmosphere.

The fallacy that forecasters should be able to get the forecast right every time, however, is widespread. Weather forecasting is one place where science falls to intense daily scrutiny, not just by the public, but also the media. Every day the world's foremost forecasters put themselves to the test by issuing weather forecasts. Usually they get the forecast right, and their good work tends to go unnoticed. However, it is when they get the forecast wrong that they get noticed, and in the shadow of media scorn poured in the direction of national weather agencies, the independent forecasting companies seem to shine. Public criticism of incorrect weather forecasts is certainly not a new thing, either – even in Fitzroy's day, the media were critical. His early forecasts, which used only simple, statistical approaches, were generally ignored by the public in favour of the fact that he sometimes gave incorrect forecasts – criticism that Fitzroy took very personally.

The reason why some forecasts go wrong is the **chaotic** way that the atmosphere behaves. It does obey a strict set of physical rules, but, as the atmosphere is so vast, a small change in the state of the atmosphere could eventually grow and have extensive impacts. American meteorologist Edward Lorenz popularised the principle of the **butterfly effect**, where the flap of a butterfly's wings in one part of the world could eventually lead, for example, to a tornado in a different part. In other words, any errors in either our initial conditions or the mathematical equations we use to project our weather forward in time could rapidly snowball into more significant errors that can affect the weather patterns on a much larger scale. This means that, as time goes on through a forecast run, the accuracy of the forecast will decrease. Over the first few days, forecasts can give reliable depictions of the weather, but by seven days into the run they are rarely likely to give an exact depiction of the weather – although they may still be able to give an idea of the overall weather situation. However, sometimes forecasts of only a day ahead can go badly wrong – particularly if a very rare weather event happens that is neither captured well by the model, nor seen before by the forecaster. It is therefore crucial, if we are to make accurate forecasts, for the assumptions we make in our model to be as realistic and physically based as possible.

Despite this, the world of forecasting is certainly not all doom and gloom. While we often hear all about wrong forecasts, the fact remains that modern forecasting techniques and advances in supercomputer technology have made our forecasts more accurate than ever. Over the first few days of a forecast run, we can be pretty confident that the model will predict correct weather conditions; over the next few days, the models are usually good at determining the overall weather pattern. Research into the development of improved forecasting techniques and new ways of representing various atmospheric effects is constantly ongoing – and with this, while we accept that no weather model will be able to predict the weather all the time, our success rate looks set to increase.

16 The Changing Climate

No book about the weather would be complete without a chapter on the issues surrounding the changing **climate**. Over the last few decades, the topic of climate change and global warming has become as much a part of our daily lives as weather forecasts. In the same way that we all have our own opinions about the weather and how good recent forecasts have been, most of us now have a view on global warming. These opinions are fuelled by the increased coverage of the topic of climate change in the news. Stories surrounding the latest discoveries concerning the climate frequently appear in the newspapers nowadays, and the media is often the first to attribute a natural disaster somewhere in the world to global warming.

This makes climate change a hot topic for debate. We are bombarded with stories linking global warming to carbon dioxide emissions, yet hear conflicting stories from those who believe that the science behind global warming is unfounded. However, climate change is a global phenomenon that will have effects on every part of the world and is certainly not something we can just ignore.

16.1 Past Records of Climate

While meteorology, the study of weather, is mainly concerned with predicting and observing the weather over a relatively short period, climatology concerns itself with long-term trends and averages in weather – perhaps most notably, trends in global mean temperature. To identify these trends, we require observations of weather over a long period. We have already seen that mankind has been observing the weather for thousands of years. However, only the last 400 years or so of weather observation have been quantitative and records are only this long in a few locations. Modern observations provide very full pictures of the atmosphere, and using archived data from the observation network can provide a reliable picture of recent weather conditions over the last 50 years or so. However, before the era of satellites, radiosondes and radar, only surface observations were available, with data quality being an issue before observations were routinely recorded in a standard format.

Collating the observational data from such a wide range of sources to produce a time series of global mean temperature is the first challenge. Several institutions around the world attempt this and produce their own trends of observed temperature stretching back over the last few hundred years – the data from the Climate Research Unit at the University of East Anglia, UK, is shown in Figure 16.1. Care is needed to ensure that dubious measurements are excluded and any measurement uncertainties are taken into account. Changes in measurement technique or instrumentation may lead to artificial, abrupt changes in the data, which must also be accounted for. A common message emerges from all of these global mean temperature time series: from year to year, there are seemingly random

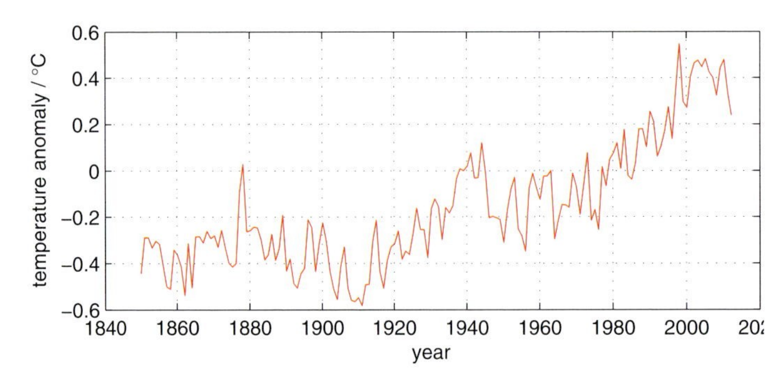

Figure 16.1 Global mean temperature time series, derived from observations made over the last 160 years. (Adapted from Brohan *et al.*, 2006.)

fluctuations in annual average temperature, yet there is an underlying trend of warming through the twentieth century.

There is evidence that the Earth's climate has always been changing. Just as the weather naturally changes from day to day, the climate can naturally change from decade to decade, century to century, millennium to millennium and beyond. Information about the Earth's temperature in the past can be gleaned from a number of natural sources. Analysis of tree rings, or **dendrochronology**, can give indications of weather conditions over the lifespan of a tree (up to maybe a few hundred years ago). Colder weather can put the tree under stress, hence it grows more slowly and its rings become closer together. While not a direct indicator of temperature, tree ring analysis can be used to give an idea of what past temperature was like – referred to as a **temperature proxy**.

Other sources can provide such temperature proxies. In Greenland and Antarctica, when snow falls, it is rarely warm enough for it to melt. As a result, it builds up, pressing the snow beneath into layers of ice, with the age of the ice increasing with depth. Extracting a core of this ice using a hollow drill can provide us with information over the course of the last 100,000 years or so (Fig. 16.2). One ice core from Antarctica contains data extending back as far as 800,000 years ago. The nature of the ice (density and colour) can give an idea of the temperature at the time, although isotopic analysis of the oxygen molecules within the ice gives more detailed information. Gas bubbles trapped within the ice are like tiny time capsules of the composition of the atmosphere.

Sediment cores can provide even longer temperature records. Ocean sediment consists of a combination of mineral particles and dead creatures that accumulate on the ocean floor; again, gradually building up over time. These can give information over much longer time periods (possibly millions of years). Analysing the presence of different types of fossilised creatures is one of many ways touse sediment cores to give an idea of temperature.

Again, collating all the available climatological information into a continuous record of global temperature is a challenge which must take into account not only the uncertainties in the temperature proxies, but also the fact that all three of these sources rely on data taken from an individual point. The result is a temperature time series that shows fluctuations on a wide range of scales, with colder

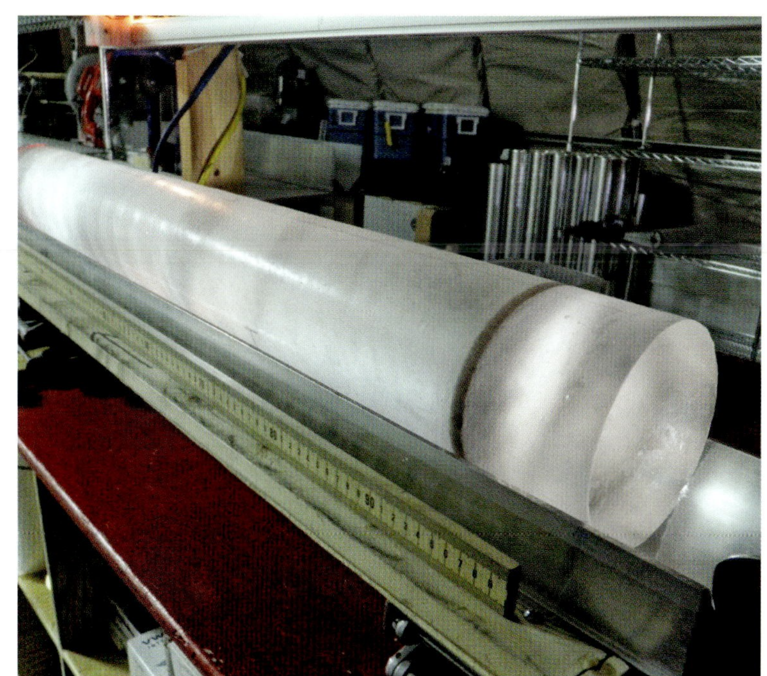

Figure 16.2 Information about the past climate can be gained from ice cores. (Photo: National Science Foundation/Heidi Roop.)

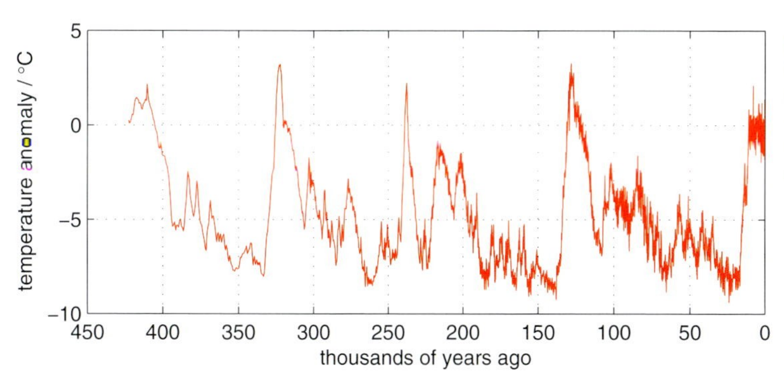

Figure 16.3 Global mean temperature series over the past 420,000 years, derived from ice core data in Vostok, Antarctica. (Adapted from Petit *et al.*, 1999.)

periods, known as **ice ages**, interspersed with periods where temperatures have been markedly warmer, as seen in Figure 16.3. Within the last million years, it is estimated that the global mean temperature varied between 10°C colder and 5°C warmer than it is today, with alternating warm and cold periods lasting a few tens of thousands of years. The last 12,000 years (defined by geologists as the **Holocene** epoch) have been marked by fairly steady global temperatures, with fluctuations of a few tenths of a degree that last a few centuries. The colder periods of this cycle are referred to as **little ice ages**. However, during the twentieth century, the global mean temperature has risen by about 0.8°C – this is likely to be a much greater change than at any other time during the rest of the Holocene.

16.2 Increasing Greenhouse Gas Concentrations

In all our records of past climate, there have been very few instances where global mean temperature has changed as markedly as it has recently. The first step to understanding why this change has occurred is to identify the primary cause. It is well known (and has been for over several decades) that human activity has changed the composition of the atmosphere. Since the Industrial Revolution, the burning of fossil fuels (oil, coal and natural gas) has released vast amounts of carbon dioxide (CO_2) into the troposphere, and growth in agriculture has increased concentrations of atmospheric methane (CH_4) – both of which are greenhouse gases. **Greenhouse gases** are molecules that exist in the atmosphere – many of which occur naturally – that, on account of their strong absorption and emission of thermal radiation, act to keep the Earth's surface warm – possibly as much as 30°C warmer than it would be in their absence (see Chapter 7). Recently, changes in concentrations of these greenhouse gases have been monitored in detail. The site at Mauna Loa Observatory in Hawaii has recorded a rise in carbon dioxide concentrations from about 320 parts per million in 1960 to over 380 parts per million today (Fig. 16.4).

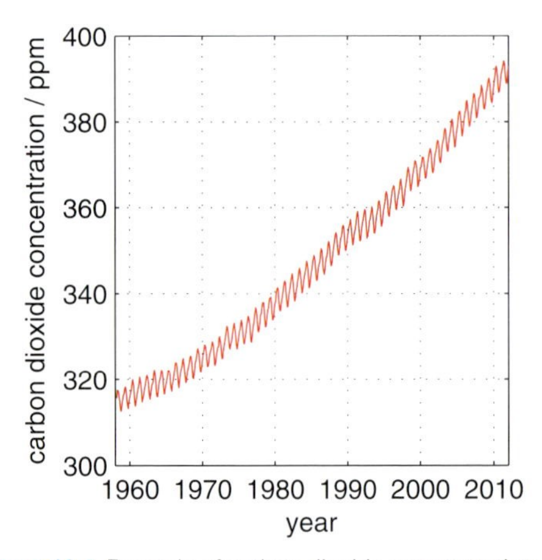

Figure 16.4 Records of carbon dioxide concentrations from Mauna Loa, Hawaii. (Data: Pieter Tans and Ralph Keeling.)

The first connection between changing carbon dioxide concentrations in the atmosphere and changing global temperature was made in 1896 by Swedish scientist Svante Arrhenius. With the onset of the Industrial Revolution already over a century old, global carbon dioxide concentrations had certainly increased. In 1938, Guy Callendar went on to tentatively suggest that this excess carbon dioxide in the atmosphere might lead to substantial changes to global climate. The big step forward in understanding the Earth's climate was brought about by the use of computer models that could encapsulate our understanding of the physics of the climate system. The first investigations into global warming using computer models were performed by a group in the USA headed by Japanese meteorologist Syukuro Manabe. He created a computer model that showed a link between carbon dioxide emissions and temperature rise as early as 1967, and became the first to run a weather model to simulate climate. He investigated the effect of an instantaneous doubling in carbon dioxide concentrations, and found a marked increase in global surface temperatures.

The simple message that has been emanating from the world of climate research for the last few decades is that the world is becoming markedly warmer, and that anthropogenic emissions of greenhouse gases, most notably carbon dioxide, are likely to be to blame. In 1988, the World Meteorological Organization founded the **Intergovernmental Panel on Climate Change**, or IPCC. It had three main goals: to investigate the causes of climate change, to review the potential impacts, and to look into ways we can deal with these impacts. Since then it has put together four Assessment Reports, written by a team of climate scientists. Each report draws together and reviews all the recent findings from climate scientists around the world and summarises them in a report on the current state of our understanding of the climate.

In its First Assessment Report of 1990, the IPCC stated that increasing carbon dioxide was likely to cause a global warming effect, but that this effect would not be discernible for ten years. The subsequent IPCC Assessment Reports have become increasingly confident about the link between anthropogenic carbon dioxide emissions and global warming. The Second Assessment Report of 1995 claimed evidence of this link; the Third Assessment Report of 2001 stated that there was 'stronger evidence' of the link. By this time, the potential threat of climate change was well recognised. The IPCC's most recent report, the Fourth Assessment Report of 2007, stated that the existence of global warming was 'unequivocal' and that most of the warming effect since the middle of the twentieth century was 'very likely' caused by anthropogenic emissions of greenhouse gases. The next update, the Fifth Assessment Report, is due for publication in 2014.

16.3 Climate Models

Modern climate models are the most important tools in both understanding the past climate and predicting the future climate. Indeed, climate simulations using these models form the backbone of the IPCC's reports. In essence, the most sophisticated climate models are the same as weather forecast models. However, many modifications are required to allow a forecast model to perform a climate simulation. For a start, we want a climate model to run over a much longer period than a weather forecast – weather forecast models simulate a week or so; climate runs simulate decades or even centuries. Simulating 100 years of climate using a model designed to produce weather forecasts would take over a year. So we use a coarser grid with larger gridboxes and fewer layers in combination with a longer time step, meaning that a single day of model simulation requires many fewer calculations. A typical gridbox size for a climate model is 200 km to 250 km. At this resolution, the climate model may struggle to predict localised effects, but can still represent both global and regional changes to the state of the atmosphere.

The atmosphere is actually only a small part of the climate system. We usually think of the weather as an atmospheric phenomenon but, on climatological timescales, changes in weather can have many effects on other parts of the Earth system, such as the oceans, sea ice and the biosphere. If we are to fully model the changing climate, we therefore need to be able to represent how changes to the average state of the atmosphere impact, for example, circulations within the oceans or the distribution of vegetation – both of which are unlikely to change much during a 10-day weather forecast, but could change drastically over the course of a 100-year climate simulation.

Modelling all of these interactions within the Earth system presents itself as the major challenge to climate science. The situation is further complicated by the existence of so-called **climate feedbacks**, where changes to the state of the atmosphere affect another element of the climate system, which in turn has a further effect on the state of the atmosphere. For example, one of the observed effects of global warming (indeed, one of the main indicators used by the IPCC to show that it is happening) is the decrease in extent of ice and snow cover in the Arctic over the past few decades. There is a large difference between the albedo of the snow and a land surface (see Chapter 7). The loss of snow will therefore decrease the Earth's average albedo, resulting in less solar radiation being reflected back to space, and hence further heating the Earth–atmosphere system. In other words, the loss of snow is acting to enhance the effects of global warming in a **positive feedback** process.

This is just one of a myriad of ways that different parts of the Earth–atmosphere system influence each other under the effects of global warming. Not all of the feedbacks are positive – some changes brought about by global warming act to reduce the warming effect via **negative feedback** processes. To capture as many of these feedbacks as we can, we must include numerical representations of as many of these processes as possible. For a start, an atmospheric climate model needs to be coupled to a full ocean model. Vast amounts of heat can be stored in the oceans, with changes in temperature and salinity both potentially affecting ocean circulation, which can cause marked changes to regional climate. The model must also be able to represent snow and ice cover. Clouds and aerosols also play an important part in climate simulations, although the feedback processes involved with these are very complicated and their understanding is currently incomplete and an area of extensive research.

16.4 Simulating Past and Future Climate

So how do we go about generating a simulation of climate conditions using our model? Before we start thinking about that, we need to validate our climate model – in other words, check that it is capable of producing realistic, sensible results over the historical period. This **validation** process is performed by running the model using climate data from the past. To perform this validation, we need to feed the model with past trends in various **climate forcings**. Forcings are external impacts that perturb the climate system by creating an imbalance in energy entering and leaving the Earth–atmosphere system, ultimately leading to a change in global temperature. Climate forcings include not just greenhouse gas emissions, but emissions of aerosols, the effect of aviation on the upper air via both exhaust gases and contrails, and many more. Our understanding of many of these forcings remains poor (for example, the effects of aerosol on global temperature) and are currently the subject of extensive research.

Natural forcings include variation in total solar irradiance and sulphate emissions from volcanic eruptions. Two climatically significant volcanic eruptions have occurred during our records for the past 50 years: those of El Chichón in Mexico in 1982, and Pinatubo in the Philippines in 1991. Their explosive eruptions blasted large volumes of sulphur dioxide and sulphate aerosols into the stratosphere, which block out solar radiation and result in a temporary global cooling. A cooling of about 0.5°C was noted for a year or so after Pinatubo's eruption. Current trends in total solar irradiance measurements made during the last 30 years or so show a clear approximately 11-year cycle, which has a small climate effect, and possibly a slight upward trend, although there is a great deal of uncertainty on whether this trend is significant.

Once we have built up confidence in our model's ability to reproduce well-observed aspects of past climate, we can move on to climate change **attribution** – that is, linking the changes in global surface temperature to various climate forcings. By switching off a particular forcing, we are able to identify the change in temperature that is due to that forcing. We can also identify the climate **sensitivity** of the forcing by determining the change in global surface temperature. A forcing with a high climate sensitivity causes larger changes in global surface temperature for a small change to radiation budget; one with a low climate sensitivity causes a small change in global surface temperature for a larger change in radiation budget.

Next, we can start to think about simulating future climate. However, this presents us with our next challenge. To create predictions of future global temperature, we first require forecasts of how greenhouse gas emissions will vary with time over the next century. This task is way beyond the remit of climate scientists, as our greenhouse gas emissions are likely to be influenced by economical, political and sociological change. For example, the development of new, efficient methods of generating power on a large scale without the need for fossil-fuel burning might gradually reduce emissions, yet the rapidly increasing global population and its improving standard of living would increase the demand for energy and resources, possibly increasing emissions. This adds an extra degree of uncertainty to the remaining scientific uncertainty.

As there is so much uncertainty in what the future could hold for greenhouse gas emissions, the IPCC have devised **future climate scenarios**. The scenarios contain possible trends of how greenhouse gases might vary in the future for different assumptions of global economics, population growth and political situations. By running a number of climate models at weather institutions around the world using these predicted changes in greenhouse gases, we can then get an idea of how global temperature might change for each scenario. The results for six scenarios have been published in the IPCC's Fourth Assessment Report (Fig. 16.5). All of their climate scenarios show an overall warming by the end of the twenty-first century, even though in many of the scenarios the greenhouse gas concentrations by then are declining. The 'B1' scenario, which depicts global movement towards environmental sustainability, predicts a warming in global temperature of between 1.1°C and 2.9°C between the period 1980 to 1999 and the 2090s. In contrast, for the 'A1FI' scenario, in which there is a global movement towards fossil-fuel-intensive economic growth, the warming is predicted to be in the range 2.4°C to 6.4°C.

Research into modelling the climate is an ongoing process, with climate scientists at national weather services and universities working hard to develop, improve and use the

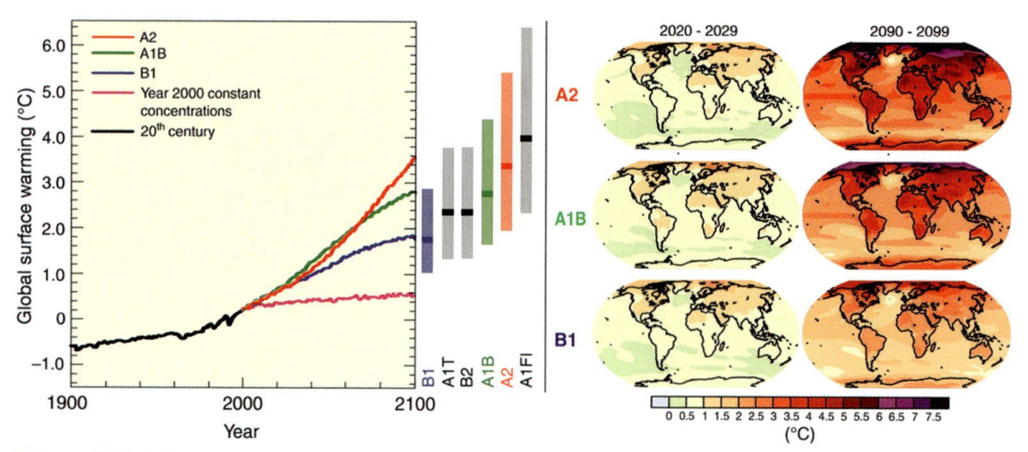

Figure 16.5 Left panel: solid lines are multi-model global averages of surface warming (relative to 1980–1999) for the SRES scenarios A2, A1B and B1, shown as continuations of the 20th century simulations. The orange line is for the experiment where concentrations were held constant at year 2000 values. The bars in the middle of the figure indicate the best estimate (solid line within each bar) and the likely range assessed for the six SRES marker scenarios at 2090–2099 relative to 1980–1999. The assessment of the best estimate and likely ranges in the bars includes the Atmosphere–Ocean General Circulation Models (AOGCMs) in the left part of the figure, as well as results from a hierarchy of independent models and observational constraints. Right panels: projected surface temperature changes for the early and late 21st century relative to the period 1980–1999. The panels show the multi-AOGCM average projections for the A2 (top), A1B (middle) and B1 (bottom) SRES scenarios averaged over decades 2020–2029 (left) and 2090-2099 (right). (Figure: IPCC Fourth Assessment Report.)

next generation of climate models in preparation for the IPCC's Fifth Assessment Report. Not all studies focus on global climate change: many focus on regional impacts. Others focus on the effects of climate change on individual aspects of the Earth system, such as the biosphere. We also constantly work to improve our scientific understanding of some of the trickier aspects of the climate system; for example, by dedicated field campaigns, satellite measurements and the development of new theories.

16.5 Adaptation versus Mitigation

The rising temperature associated with greenhouse gas emissions is, of course, only a small part of the global warming story. The rising temperatures look set to impact many other aspects of the Earth–atmosphere system, possibly leading to hazardous consequences. For a start, melting ice in the polar regions, in combination with the thermal expansion of the ocean, could lead to the rise of sea levels – the IPCC report predicts rises of between 0.2 m and 0.6 m across their six scenarios by the 2090s. This is particularly hazardous for inhabited coastal areas. Many of the world's largest cities lie near the coast a few metres above sea level and are potentially at risk. Indeed, some Pacific islands, barely extending a few metres out of the sea, risk inundation.

Changes in weather patterns are likely to have an effect on agriculture. In some regions of the world, this is likely to be a positive effect – warmer temperatures in mid-latitudes could well increase productivity and allow farmers to grow during times of the year when they might not have been able to in the past. In cold regions such as Siberia, increasing temperatures may allow farming to spread into areas where it may not have been viable before. Warmer temperatures in the tropics, however, are likely to decrease agricultural productivity as temperatures become too hot for crops to grow. In poorer parts of the world, loss of crops could cause widespread famine and malnutrition. And it is unlikely to be only humans that feel the effects of global warming: ecosystems around the world could be drastically affected by rising temperatures. Some species may well thrive in warmer conditions, while others face potential extinction.

One approach to dealing with climate change is to accept that it is happening and modify our lifestyles to deal with it – the principle of **adaptation**. It may not be the best solution in the interests of the planet and may not save species from extinction. But we can, for example, build higher sea defences to protect ourselves from rising sea levels. We can change our farming habits, developing and growing hardier crops, and come up with systems of irrigation. The downside of a global principle of adaptation is that it is not a global solution – some parts of the world may not, for example, have enough money to build defences against the sea.

An alternative is to tackle the problem of global warming head-on via **mitigation**. This involves taking action to reduce and, if possible, reverse the effects of global warming. However, for mitigation to happen, international co-operation is required. In 1992, the Earth Summit was held in Rio de Janeiro, which pulled together leaders and politicians from 172 governments to discuss a number of issues threatening the Earth, including the reduction of the burning of fossil fuels. It led to the creation of the Kyoto Protocol in 1997 – an international agreement setting countries targets for greenhouse emissions in the period between 2008 and 2012 with the aim of stabilising greenhouse gas concentrations in the atmosphere. Most of the countries of the world have now ratified the Protocol, with the exception of the USA. However, while a great many countries have limited their emissions in accordance with the Protocol, few have actually reached their targets.

Recently, even more radical approaches to mitigation of climate change have been proposed. Some scientists have suggested the principle of **geoengineering** – in other words, modifying the planet in some way to counteract the warming caused by the excess greenhouse gases. Suggestions include the launching of millions of mirrors into space to reflect a fraction of direct sunlight, hence reducing the total solar irradiance, or the injecting of sulphates into the stratosphere to create a cooling similar to that seen after a volcanic eruption. But a great deal of further research would be required before any such measure could be undertaken, to ensure no detrimental side effects.

Mitigation of climate change still remains a hot topic today, with political negotiations for a successor to the Kyoto Protocol still ongoing. However, for the future, mitigation alone is not the answer to climate change. Whatever happens, it is likely that the current warming trend will continue, even if we could instantaneously stop emitting carbon dioxide. In other words, while we are making efforts to mitigate climate change, we need to bear in mind that the climate will continue to change, and that we will need to adapt to it.

Glossary

A

absolute zero [8]**:** the lowest theoretically possible temperature: –273.15°C or 0 K.

adaptation [135]**:** an approach to dealing with the changing climate by modifying our lifestyles.

adiabatic [57]**:** an adiabatic process involves no transfer of heat between a system and its surroundings.

advection fog [59]**:** fog generated by the transport of warm, moist air over a colder surface, which causes the air to cool below its dew point, and condensation to occur.

aerosols [43]**:** small particles that are suspended in the atmosphere. Aerosols include dust, soot and sea salt.

African Easterly Jet [88]**:** a low-level jet stream that flows east to west off the coast of western Africa and plays a part in the formation of Atlantic hurricanes.

ageostrophic wind [67]**:** the component of the wind that flows in a direction parallel to the pressure gradient (perpendicular to the isobars).

air frost [17]**:** an air frost is said to occur if the air temperature falls below 0°C.

air mass [75]**:** a large volume of air having fairly uniform properties, such as temperature and humidity.

albedo [49]**:** the fraction of incoming radiation that is reflected from a surface.

altocumulus [23]**:** a mid-level cloud, usually made up of a sheet of many small cells or rolls of cloud.

altocumulus lenticularis [105]**:** lens-shaped clouds that form downwind of hills and mountains.

altostratus [23]**:** a mid-level cloud, consisting of a fairly uniform layer. Altostratus may be thick enough to obscure the Sun.

anabatic wind [107]**:** an uphill flow of air on a mountainside. The warming of air on a mountainside causes the air to become less dense than its surroundings and flow uphill.

analysis [113]**:** a global snapshot of weather conditions that is used to initialise a forecast run.

anemometer [18]**:** a term for any device that measures wind speed.

aneroid barometer [18]**:** a type of barometer that uses a partially evacuated metal capsule to measure changes in air pressure.

anomalous propagation [29]**:** false returns on a radar image caused by refraction of the radar beam under a temperature inversion.

anticyclone [74]**:** an area of high pressure.

anticyclonic gloom [75]**:** long periods of overcast weather associated with cloud trapped in an anticyclone.

anvil cloud [96]**:** the vast shield of ice cloud that spreads out around the top of a cumulonimbus.

aphelion [62]**:** the point of the Earth's orbit when it is furthest from the Sun.

Arctic air mass [76]**:** a mass of air that originated over the Arctic Ocean.

argon (Ar) [34]**:** an inert gas that makes up nearly 1% of the Earth's atmosphere.

atmosphere [33]**:** the narrow layer of air that enshrouds the Earth.

atmospheric window [50]**:** a range of wavelengths in the electromagnetic spectrum (between 8 μm and 11 μm) in which thermal radiation is largely transmitted through the atmosphere.

attenuation [29]**:** reduction of strength of a radar beam as it passes through heavy rain.

attribution [133]**:** the linking of changes in global temperatures to various climate forcings.

Azores High [71]**:** the area of high pressure that is often found over the sub-tropical Atlantic.

B

backing [18]**:** the wind is said to be backing if it is changing in an anticlockwise direction (for example, from northerly to westerly).

barograph [19]**:** a barometer that keeps track of air pressure over time, allowing trends in pressure to be determined.

barometer [6]**:** a term for any device that measures air pressure.

Beaufort scale [8]**:** a scale of wind speed categories that runs from 0 (calm conditions) to 12 (hurricane conditions), developed by Francis Beaufort.

Bergeron–Findeisen process [44]**:** a process by which ice crystals in a volume of supercooled liquid water droplets can rapidly grow at the expense of the liquid water.

black bodies [48]**:** theoretical objects that emit electromagnetic radiation with maximum efficiency.

blocking anticyclone [78]**:** an anticyclone that remains in location for long periods of time and disturbs the passage of mid-latitude depressions.

bomb [82]**:** a term used to describe any mid-latitude depression whose central pressure falls by more than 24 hPa in 24 hours.

boundary conditions [117]**:** information about the weather conditions at the boundaries of a smaller-scale model that must be fed in from a larger-scale model to keep the regional model running.

boundary layer [38, 116]**:** the lowest part of the troposphere where the flow is affected by interaction with the Earth's surface.

butterfly effect [126]**:** attributed to Edward Lorenz, an analogy used to describe the chaotic nature of the atmosphere: that the flap of a butterfly's wings in one place could grow to initiate storms in another.

Buys Ballot's law [66]**:** this states that, if an observer backs the wind in the northern hemisphere, low pressure is to the observer's left.

C

Campbell–Stokes sunshine recorder [20]**:** a device for measuring sunshine duration, consisting of a glass orb that focuses sunlight onto a strip of card.

cap cloud [106]**:** a smooth cloud that forms on the top of a hill or mountain.

carbon dioxide (CO_2) [34]**:** a greenhouse gas that exists in the atmosphere in small quantities, but is an important factor in global warming.

Celsius scale [7]**:** a temperature scale first defined by Anders Celsius, which uses the freezing point of water (0°C) and the boiling point of water (100°C) as reference points.

Central England Temperature [14]**:** a temperature record from stations in central England that stretches back to 1659 – the longest existing observational record.

chaotic [126]**:** a chaotic system follows a strict set of rules, but has so many degrees of freedom that the smallest change at one location within the system could have large impacts on the state of the system after a period of time.

cirrocumulus [23]**:** a high-level ice cloud made up of small, shadowless cells or ripples of cloud.

cirrostratus [23]**:** a high-level, thin and tenuous sheet of ice cloud that makes the sky look pale and can produce optical effects around the Sun.

cirrus [21]**:** high-level, wispy streaks of ice cloud.

climate [127]**:** the average weather conditions in a location over a long period of time.

climate feedback [132]**:** any process that causes a change in the state of the atmosphere where the change then acts to either strengthen or weaken the process.

climate forcing [132]**:** an impact that perturbs the balance of the climate system.

climate modelling [13]**:** the act of using modified weather forecast models to predict conditions over much longer periods of time.

climatic observations [15]**:** a set of observations that are only made once a day.

cloud cover [11]**:** the fraction of the sky that is obscured by cloud, usually quoted in eighths or oktas.

cloud head [81]**:** a comma-shaped hook of cloud found in the centre of a severe mid-latitude depression, usually associated with heavy rain.

clustering [119]**:** the division of an ensemble forecast into groups of forecasts that show similar predictions of weather conditions.

coalescence [43]**:** the growth of raindrops by individual drops colliding and merging with other drops.

cold conveyor belt [81]**:** one of the three air streams that maintains a mid-latitude depression, consisting of a stream of cold air that flows into the centre of the depression from ahead of the warm front.

cold-core anticyclone [75]**:** a type of anticyclone driven by surface cooling, which leads to increased air density near the surface.

cold desert [108]**:** the sort of desert found in the Arctic and Antarctic, which usually have icy surfaces.

cold front [76]**:** a boundary where a mass of cold air undercuts a mass of warmer air, leading to ascent and cloud development.

condensation [7]**:** the conversion of water vapour into liquid water.

conduction [59]**:** the transfer of heat through a solid object by molecular vibrations.

continuity [65]**:** the principle that, in a flow, fluid cannot be created or destroyed.

convection [56]**:** the transfer of heat within a fluid by motion of the fluid, which leads to contrasts in properties within the fluid becoming smoothed out.

convective clouds [58]**:** a generic term for clouds that form when parcels of warm, moist air rise vertically in unstable conditions and saturate. Convective clouds include cumulus and cumulonimbus.

convective overshoot [96]**:** a dome found on the top of a cumulonimbus above the updraught, where the ascending air is pushing against the tropopause.

convergence [65]**:** this occurs at a point in a fluid when there is more horizontal flow into the point than there is leaving it.

Coriolis force [66]**:** the effective force on a mass of air that is caused by the rotation of the Earth.

cumulonimbus [21]**:** tall, deep clouds consisting of liquid water near their bases and vast shields of ice at the top. These clouds bring showers of rain, hail, thunder and lightning.

cumulus [21]**:** low-level clouds that have a flat base and extend upwards in a dome.

cumulus congestus [21, 95]**:** tall cumulus clouds that may be deep enough to bring some rain.

cumulus humilis [21, 94]**:** small, flat cumulus clouds.

cyclogenesis [76]**:** the formation and strengthening of mid-latitude depressions.

D

data assimilation [115]**:** the technique of mathematically combining weather forecast model output with observations such that the resulting state of the atmosphere contains no unrealistic artefacts.

dendrochronology [128]**:** the analysis of tree rings; used to inferpast climate conditions.

density [35]**:** the mass of a substance per unit volume, usually expressed in kg m^{-3}.

depression [74]**:** an area of low pressure.

deterministic forecasting [119]**:** a deterministic weather forecast uses a single model run that gives a single prediction of how the weather may evolve.

development site [78]**:** a location where conditions are favourable for the development of mid-latitude depressions.

dew [59]**:** liquid water that condenses on the surface when air adjacent to the surface is cooled to its dewpoint temperature.

dewpoint temperature [42, 59]**:** the temperature to which air must be cooled to cause it to saturate, assuming no changes to its pressure and its specific humidity.

diffuse radiation [50]**:** light from the Sun that has reached the Earth's surface having been scattered at least once by the atmosphere or its contents.

diffusion [43]**:** the tendency of water vapour to flow from an area of higher concentration to an area of lower concentration.

direct radiation [50]**:** light from the Sun that has reached the Earth's surface without having been scattered.

discretisation [116]**:** the mapping of a continuous field of data onto a grid at fixed points.

dispersion [53]**:** the spreading out of light into its different wavelengths.

divergence [65]**:** this occurs at a point in a fluid when there is more horizontal flow leaving the point than there is arriving at it.

domain [117]**:** the area over which a weather forecast model runs.

Doppler radar [30]**:** a type of radar that transmits pairs of pulses to allow the movement of areas of rainfall to be determined.

downdraught [97]**:** the descending column of air at the heart of a convective storm system.

dry adiabatic lapse rate [57]**:** the rate at which the temperature of a parcel of air will fall per unit height if it rises adiabatically and the air does not saturate.

dry-bulb thermometer [16]**:** a standard thermometer with its bulb exposed to the air, used to measure air temperature.

dry conveyor belt [81]**:** one of the three air streams that maintains a mid-latitude depression, consisting of a flow of dry air into a mid-latitude depression behind the cold front.

dry intrusion [81]**:** a cloudless band of drier air in a mid-latitude depression associated with the dry conveyor belt.

dust devil [108]**:** a small vortex extending upwards from the ground, visible when it picks up dust.

E

electromagnetic radiation [8, 48]**:** all radiation that is caused by oscillations of electric and magnetic fields. In meteorology, important types of electromagnetic radiation are visible light, ultraviolet and infrared.

El Niño [92]**:** a warming of the sea surface of the eastern Pacific Ocean, which has far-reaching impacts on world weather.

emissivity [49]**:** the efficiency at which a surface emits radiation with respect to a black body at the same temperature.

Enhanced Fujita–Pearson Scale [101]**:** the scale on which the severity of tornadoes is quantified.

ensemble forecasting [118]**:** the act of running a number of forecasts with slightly different initial conditions to give an idea of forecast uncertainty.

environmental lapse rate [57. 94]**:** the background rate of change of temperature of the atmosphere with height.

equilibrium [49]**:** a system in equilibrium is in perfect balance: for example, it gains the same amount of energy as it loses.

evaporation [41]**:** the conversion of liquid water into water vapour.

eye [90]**:** the small patch of clear sky found in the centre of a tropical cyclone caused by local subsidence of the air.

eye wall [90]**:** the band of storm cloud surrounding the eye of a tropical cyclone.

F

Fahrenheit scale [7]**:** a temperature scale, proposed by Gabriel Fahrenheit, that originally used the freezing point of water (32°F) and human body temperature (96°F) as reference points, although it now takes its reference points from the Celsius scale.

fair-weather cumulus [94]**:** another term for cumulus humilis – small cumulus clouds.

Ferrel cell [70]**:** part of the three-cell model of atmospheric circulation. The Ferrel cell is a weak circulation that spans the gap between the Hadley and polar cells and rotates in the opposite direction, with rising air in the mid-latitudes and descent over the sub-tropics.

fluvial flooding [106]**:** a type of flooding where excess rainfall causes the ground to become saturated and water to run straight off the surface.

fog [58]**:** cloud that forms at the surface and reduces visibility to less than 1 km.

Föhn effect [107]**:** an effect that creates warm, dry winds in the lee of mountain ranges.

forked lightning [98]**:** lightning that passes from cloud to ground, allowing its branching structure to be seen.

freezing fog [59]**:** fog, consisting of supercooled liquid droplets that instantly freeze onto any cold surface.

freezing rain [59]**:** raindrops that are composed of supercooled water, which freeze onto cold surfaces leading to the build up of crusts of ice.

frontal wave [76]**:** a wave in the polar front which marks the first stage of development of a mid-latitude depression.

frost [59]**:** ice crystals that freeze on the surface when air adjacent to the surface is cooled to its dewpoint temperature and the temperature is cold enough.

funnel cloud [101]**:** a funnel-shaped protrusion from the base of a cloud. Funnel clouds that extend to the ground become tornadoes.

future climate scenarios [133]**:** trends of how greenhouse gas emissions may continue into the future, devised for the IPCC Assessment reports.

G

Galileo thermometer [6]**:** a type of thermometer consisting of a number of glass spheres suspended in a fluid within a closed glass container.

geoengineering [135]**:** a number of proposed projects to tackle increasing global temperatures by modifying the planet.

geostationary satellite [30]**:** a satellite in geostationary orbit travels directly above the Equator and takes exactly one day to travel around the Earth once. This means that it always remains in the same position in the sky.

geostrophic wind [67]**:** the component of the wind that flows in a direction perpendicular to the pressure gradient (parallel to the isobars).

global model [117]**:** a weather forecast model that generates forecasts on a grid that covers the whole world.

grass minimum temperature [17]**:** the lowest temperature reached overnight, measured using a thermometer at grass level.

graupel [46]**:** small, soft ice particles a few millimetres in diameter.

Gravity [36]**:** the downward force felt on Earth that is due to the Earth's mass.

greenhouse gases [50, 130]**:** a number of gases present in the atmosphere, including carbon dioxide, water vapour and methane, which absorb thermal radiation and warm the Earth.

Greenwich Mean Time [14]**:** the time in London (excluding daylight saving time), used to determine UTC.

ground clutter [29]**:** false returns on a radar image caused by reflection of the radar beam off the ground and other immovable objects.

ground frost [17]**:** a ground frost is said to occur if the grass minimum temperature falls below 0°C.

guidance [120]**:** the first step in issuing a forecast. The guidance is an overview of how the weather is expected to develop and is compiled by senior forecasters.

gust front [97]**:** the boundary of the plume of cold air from the downdraught of a cumulonimbus that spreads out when it reaches the ground.

H

haboob [108]**:** a wall of dust, raised by an advancing gust front ahead of a storm.

Hadley cells [68]**:** part of the three-cell model of atmospheric circulation. The Hadley cells are convective circulations that span the tropics, with their ascending branches over the Equator and their descending branches over the sub-tropics.

halo [54]**:** a circle of light refracted off ice crystals that can sometimes be seen around the Sun or Moon.

heat capacity [102]**:** a measure of how much heat energy is required to warm a surface by a certain temperature difference.

hectopascal (hPa) [18]**:** a standard unit

of pressure, equivalent to 100 pascals or 1 millibar. Standard atmospheric pressure is 1,013.25 hPa.

heterogeneous nucleation [43]**:** the condensation of liquid droplets or ice crystals with the assistance of an aerosol particle.

heterosphere [39]**:** the part of the atmosphere where gases are no longer well mixed. This extends upwards from about 100 km above the surface.

hole-punch cloud [46]**:** a hole that forms in a layer of supercooled liquid water cloud when glaciation is initiated, often by the passage of an aircraft through the layer.

Holocene [129]**:** the most recent epoch of geological time, defined as the last 12,000 years.

homogeneous nucleation [42]**:** the spontaneous condensation of liquid droplets or ice crystals in pristine, clean air.

homosphere [39]**:** the part of the atmosphere where the gases are well mixed. This extends from the surface up to a height of about 100 km.

hot desert [108]**:** the sort of desert found in the tropics and sub-tropics that has a sandy or rocky surface.

humidity [17, 41]**:** a measure of the amount of water vapour held in the atmosphere.

hurricane [88]**:** a name for intense tropical cyclones that form in the Atlantic.

hydrological cycle [41]**:** the continuous passage of water between the ocean, the atmosphere and the land.

hydrostatic balance [64]**:** for the atmosphere to be in hydrostatic balance, the upward pressure gradient force must be equal to the weight of the atmosphere above.

hygrometer [7, 17]**:** a term for any device that measures humidity.

I

ice age [129]**:** a prolonged period of colder climatic conditions, usually associated with the advance of the polar ice caps.

infrared radiation [47]**:** electromagnetic radiation with wavelengths between about 750 nm and 1 mm. Thermal radiation consists of infrared radiation.

initial conditions [14]**:** the global snapshot of information about the state of the atmosphere that is used to initialise a forecast run.

Intergovernmental Panel on Climate Change (IPCC) [131]**:** a WMO panel that summarises current research into climate change into regular Assessment Reports.

interpolation [114]**:** the act of inferring the weather conditions at a point between observations by assuming that the conditions vary smoothly.

Intertropical Convergence Zone (ITCZ) [68]**:** the zone of convergence where the Hadley cells in each hemisphere meet, generating a band of ascent marked by a patchy band of convective cloud around the tropics.

ionisation [39]**:** the charging of atmospheric molecules.

ionosphere [39]**:** a layer of charged particles found in the lower thermosphere. The ionosphere is an important part of the Earth's electric circuit.

isobars [4, 67]**:** lines on a weather map that connect places of constant atmospheric pressure.

J

jet stream [71, 73]**:** a stream of very fast-moving winds. The polar front jet stream runs in the upper troposphere in association with the polar front and is a major factor in the growth and dissipation of mid-latitude depressions.

K

katabatic wind [107]**:** a downhill flow of air on a mountainside. The cooling of air on a mountainside causes the air to become more dense than its surroundings and flow downhill.

Kelvin scale [8]**:** a scale of temperature that increases upwards from absolute zero, using the same temperature increment as the Celsius scale.

khamsin [108]**:** a warm wind that blows across the Mediterranean from Egypt to the eastern Mediterranean countries.

knot [18]**:** a unit of speed, often used in meteorological observations, equivalent to one nautical mile per hour.

L

lake effect snow [104]**:** snowfall that occurs when cold winds pass over lakes and seas, causing heating and an outbreak of convection.

laminar flow [38]**:** in a fluid, laminar flow is characterised by organised motion in layers, with no eddies or vortices.

land breeze [103]**:** a light wind that blows from the land to the sea on a clear night. It is caused by the different properties of the land and sea surfaces.

La Niña [92]**:** a cooling of the sea surface of the eastern Pacific Ocean, which has far-reaching impacts on world weather.

leader stroke [98]**:** the first stage of a lightning strike, which creates a path of ionised air.

leste, leveche [108]**:** warm winds that blows across the Mediterranean from Morocco to the Iberian peninsula.

level of neutral buoyancy [57]**:** the level at which a rising plume of air is no longer less buoyant that the surrounding air, hence it can rise no further.

lidar [29]**:** a system of cloud detection that is similar to radar, but uses light pulses instead of radio.

lifting condensation level [57]**:** the level at which air in a rising plume of air begins to condense, appearing as the level of the cloud base.

Little Ice Age [62]**:** a period of lower global mean temperature that ran from the sixteenth century to the nineteenth century.

low-Earth orbit [30]**:** an alternative term for a polar orbit.

M

Madden–Julian Oscillation [86]**:** a fluctuation in pressure that travels eastward around the tropics and alternately enhances and suppresses convection in the Intertropical Convergence Zone.

mammatus [97]**:** lumpy undulations found on the base of a cumulonimbus.

mass mixing ratio [42]**:** a method of quantifying humidity in terms of mass of water vapour per unit mass of dry air.

Maunder minimum [62]**:** a period of low solar activity between about 1650 and 1700.

maximum thermometer [16]**:** a thermometer designed to measure the maximum temperature reached over a period of time.

mean-sea-level pressure [19, 35]**:** the equivalent air pressure that would be measured at a weather station if it were situated at mean sea level.

melting layer or freezing layer [44]**:** the level in the atmosphere where the air temperature is 0°C and hence falling ice particles start to melt to liquid water.

mercury barometer [6]**:** a type of barometer that measures pressure by the height of a column of mercury.

mesocyclone [100]**:** a storm system where the updraught and downdraught spiral around one another.

mesoscale [102]**:** mesoscale weather systems are typically of size a hundred kilometres.

mesosphere [38]**:** the layer of the atmosphere above the stratosphere, marked by decreasing temperature with height.

meteogram [120]**:** a graph showing how various aspects of the weather are likely to evolve at a given point with time.

methane (CH_4) [34]**:** a greenhouse gas that exists in the atmosphere in very small quantities. It is also a major component of natural gas.

mid-latitude depression [65]**:** a structured low-pressure system that forms in the mid-latitudes, normally with rain aligned along bands called fronts.

mid-latitudes [70, 73]**:** the part of the Earth outside the Hadley circulation and containing the polar front, typically defined by latitudes between about 30° and 60°.

millibar (mb) [18]**:** a standard unit of pressure, equivalent to the hectopascal.

minimum thermometer [16]**:** a thermometer designed to measure the minimum temperature reached over a period of time.

mitigation [135]**:** an approach to dealing with the changing climate by taking action to combat it – for example, by reducing greenhouse gases.

monsoon [86]**:** the reversal of wind direction in the tropics that separates wet seasons and dry seasons. The word 'monsoon' is also sometimes applied to the heavy rain associated with the wet season.

Moore's Law [13]**:** the recent trend in computing whereby the processing power of computers has been doubling every two years.

multi-cell storm [98]**:** a convective system that consists of a number of cumulonimbus cells. These can persist for much longer than a single-cell storm.

N

negative feedback [132]**:** if a change in temperature affects a property of the atmosphere such that it reduces the temperature change, a negative feedback process is said to occur.

negative lightning [98]**:** a negative lightning strike occurs when the base of the cloud adopts a negative charge. Most lightning strikes are of this type.

Nexrad [30]**:** the network of Doppler radars in the USA.

nimbostratus [23]**:** a thick, low-level sheet of cloud that brings periods of persistent rain.

nimbus [21]**:** an obsolete term for a raincloud; one of Luke Howard's original cloud types.

nitrogen (N_2) [34]**:** a chemically inactive gas that makes up about 78% of the Earth's atmosphere.

noctilucent cloud [38]**:** thin ice clouds that form over the poles in the mesosphere.

North Atlantic Oscillation index [78]**:** an index that quantifies the position of the North Atlantic storm track, based on the strength of the north–south pressure gradient across the North Atlantic.

Norwegian model [11, 76]**:** a model describing the development and life cycle of mid-latitude depressions, developed in the 1920s by a group of Norwegian scientists including Jakob Bjerknes.

nowcasting [122]**:** the prediction of the weather for a few hours ahead by looking at radar and satellite imagery.

numerical weather prediction [112]**:** the technique of forecasting the weather using physical equations to project the state of the atmosphere at a future point in time.

O

oblate [44]**:** an oblate sphere is wider than it is tall.

occluded front [76]**:** when a cold front catches up with a warm front, it lifts it up off the surface creating an occluded front or an occlusion.

occlusion [76]**:** the process that creates an occluded front.

okta [21]**:** the standard unit of cloud cover, where one okta is one eighth of the sky.

orographic enhancement [105]**:** the intensification of rainfall caused by forced ascent over hills and mountains.

orography [105]**:** a general terms for hills and mountains.

oxygen (O_2) [34]**:** a reactive gas that makes up about 21% of the Earth's atmosphere.

ozone (O_3) [32]**:** a gas that exists in the atmosphere in small quantities but acts to shield the Earth from ultraviolet radiation from the Sun. It is found in highest concentrations in the ozone layer.

ozone depletion [51]**:** the process by which ozone in the atmosphere is irreversibly destroyed, mostly via interaction with CFCs.

ozone hole [51]**:** lower concentrations of ozone in the ozone layer over the Antarctic, particularly during spring.

ozone layer [38]**:** the layer in the stratosphere in which ozone concentrations are highest – usually found between 20 km and 30 km above the Earth's surface.

P

parameterisation [116]**:** in a weather forecast model, processes that occur on a scale smaller than that of the model grid cannot be directly represented. Instead, they must be accounted for statistically, or parameterised.

parcel of air [57]**:** a small mass of air that is assumed not to interact with the air around it – a useful model when explaining air movements.

partial pressure [42]**:** the contribution to atmospheric pressure that is due to the molecules of a particular gas.

perihelion [62]**:** the point of the Earth's orbit when it is nearest to the Sun.

photons [49]**:** a small packet of energy that travels at the speed of light. Electromagnetic radiation can be considered a stream of such packets of energy.

photosphere [47]**:** the outermost layer of the Sun, which emits light.

photosynthesis [34]**:** the process by which plants produce energy by converting carbon dioxide and water into oxygen.

pluvial flooding [106]**:** a type of flooding where excess rainfall causes more water to enter a river than it can cope with, resulting in the river bursting its banks.

polar cell [69]**:** part of the three-cell model of atmospheric circulation. The polar cell consists of rising air in the mid-latitudes at the polar front and descent over the poles.

polar continental air mass [75]**:** a mass of air that originated over land at high latitudes.

polar front [70]**:** the boundary in the mid-latitudes where warm tropical air meets cold polar air.

polar low [82]**:** a type of low-pressure system that forms at high latitudes and tracks south, usually bringing heavy snowfall.

polar maritime air mass [75]**:** a mass of air that originated over the ocean at high latitudes.

polar orbit [30]**:** a satellite in a polar orbit travels around the Earth in a north–south direction, taking about 90 minutes to orbit the Earth once.

positive feedback [132]**:** if a change in temperature affects a property of the atmosphere such that it enhances the temperature change, a positive feedback process is said to occur.

positive lightning [98]**:** a positive lightning strike occurs when the base of the cloud adopts a negative charge. They are rarer than negative lightning strikes, although they tend to be more powerful.

pressure [6]**:** atmospheric pressure is the force per unit area exerted by the atmosphere, usually expressed in pascals (Pa) or hectopascals (hPa).

pressure gradient force [64]**:** the force on air parcels caused by pressure differences within the atmosphere.

pressure trend [19]**:** any change in air pressure at a location over time.

primitive equations [116]**:** a set of basic physical equations that describe the behaviour of the atmosphere in space and time and form the basis of a weather forecast model.

probabilistic forecasting [119]**:** a probabilistic weather forecast uses many runs of a forecast model with slightly different initial conditions to give an idea of the probability of a certain weather type occurring.

profile [27]**:** the variation of a certain aspect of the weather (such as temperature) along a vertical path through the atmosphere.

prognostic chart [120]**:** a computer-generated chart, produced at various time steps through a forecast run, used to give an idea of how the weather is likely to change.

psychrometer [17]**:** a type of hygrometer that measures humidity using a dry-bulb thermometer and a wet-bulb thermometer. Humidity is determined by comparing the temperatures recorded on the two thermometers.

R

radar [11]**:** a system of rainfall detection that uses pulses of radio waves (microwaves). These reflect off raindrops, giving information about both the location and the intensity of the rainfall.

radar reflectivity [29]**:**.the strength of a reflected radar signal as a fraction of the strength of the transmitted signal.

radiation [24, 47]**:** the transfer of heat via the emission of electromagnetic waves.

radiation budget [48]**:** the balance of

incoming radiation from the Sun and outgoing radiation emitted by the Earth–atmosphere system.

radiation fog [58]**:** fog generated by cooling of the surface by emission of thermal radiation. This cools the air above to its dewpoint, initiating condensation.

radiosonde [11]**:** an electronic device that is launched on a helium balloon and remotely returns observations of the weather conditions during its ascent.

radio waves [11]**:** electromagnetic radiation with very long wavelengths, used in radio transmissions. Short-wave radio waves (microwaves) are used in radar systems.

rain shadow [107]**:** the tendency for an area downwind of mountains to receive less rainfall than an area upwind.

refraction [53]**:** the change in direction of light as it passes between two different materials (for example, air and water).

regional model [117]**:** a weather forecast model that runs on a limited area of the world at greater detail than a global model.

relative humidity [17, 42]**:** a method of quantifying humidity in terms of mass of water vapour as a fraction of the maximum mass of water vapour that the air can hold.

remote sensing [28]**:** the detection of weather conditions over large areas from a single, remote location (for example, using radar or satellites).

return stroke [98]**:** the second stage of a lightning strike, where current flows along the path of ionised air generated by the leader stroke. The return stroke is substantially brighter.

ridge [74]**:** a long extension of high pressure.

rime [59]**:** spikes of ice that freeze on a cold surface when the surface is exposed to freezing fog.

S

Saffir–Simpson Hurricane Scale [88]**:** a scale used to describe the severity of tropical cyclones.

saturated adiabatic lapse rate [57]**:** the rate at which the temperature of a parcel of air will decrease per unit height if it rises adiabatically and the air is saturated.

scattering [49]**:** the redirection of light from atmospheric molecules and particles.

sea breeze [102]**:** a light wind that blows from the sea to the land on a hot day, caused by the different properties of the land and sea surfaces.

sea breeze front [102]**:** the advancing edge of a sea breeze, where the maritime air encroaches on the air over land.

seasonal forecast [124]**:** a forecast that predicts the general weather conditions over an entire season.

seeder-feeder effect [105]**:** an effect that leads to more intense rain in hills and mountains. When rain falls through cap clouds, the raindrops grow by coalescence, resulting in much heavier rain than would fall in the absence of a cap cloud.

sensitivity [133]**:** climate sensitivity is the effect on the change in global temperature caused by a change in the magnitude of a certain climate forcing.

sheet lightning [98]**:** lightning that passes between charged regions within a thundercloud, illuminating the cloud from within.

single-cell storm [98]**:** a thunderstorm that consists of a single cumulonimbus cell.

sirocco [108]**:** a warm wind that blows across the Mediterranean from North Africa to France and Italy.

smog [110]**:** a smoky layer of fog that can form in polluted conditions.

solar radiation [48]**:** radiation that has originated from the Sun – mostly consisting of visible light and ultraviolet radiation.

solarimeter [24]**:** a device that measures the intensity of sunlight reaching the Earth's surface. It can also be used to infer sunshine duration.

sounding [27]**:** a set of measurements from a radiosonde that sample the atmosphere along a vertical path.

Southern Oscillation [91]**:** the fluctuation

of ocean temperatures between El Niño and La Niña conditions, which affects wind flow patterns and rainfall over the tropical Pacific Ocean. It also has far-reaching impacts on weather conditions elsewhere in the world.

specific humidity [42]**:** a method of quantifying humidity in terms of mass of water vapour per unit mass of air.

squall line [99]**:** an organised line of storms that can bring gusty winds and heavy showers.

stability [58]**:** a measure of how susceptible the atmosphere is to cloud formation. In an unstable atmosphere, clouds form readily; in a stable atmosphere, cloud formation is suppressed.

station pressure [18]**:** the air pressure recorded at a weather station before correction for its altitude above mean sea level.

Stevenson screen [16]**:** a ventilated box used to house thermometers and psychrometers as part of a weather station.

sting jet [82]**:** a strong, descending flow of air found in the centre of some severe mid-latitude depressions, which can cause damaging winds at the surface.

storm surge [91]**:** a rise in sea level under a severe low-pressure system, which can cause flooding in low-lying coastal areas.

storm tracks [71, 78]**:** the paths frequented by mid-latitude depressions.

stratiform clouds [58]**:** a generic term for clouds that form in sheets over large areas. These form in stable conditions, or at weather fronts, and include stratocumulus, stratus and nimbostratus.

stratocumulus [23]**:** a low-level liquid water cloud consisting of a layer of individual cells.

stratopause [38]**:** the temperature maximum found at the top of the stratosphere.

stratosphere [38]**:** the layer of the atmosphere above the troposphere, marked by increasing temperature with height and containing the ozone layer.

stratus [21]**:** a low-level liquid water cloud that has a grey and uniform appearance and sometimes brings drizzle.

street canyon [109]**:** a street bordered on each side by tall buildings, which can channel the wind in urban areas.

sub-tropical highs [68]**:** the areas of high pressure and dry conditions associated with the descending branches of the Hadley cells, found approximately 30° either side of the Equator.

sundogs or **parhelia** [54]**:** patches of dispersed light that can sometimes be seen on a horizontal plane either side of the Sun. They are caused by the refraction of light in ice crystals. Similar patches around the Moon are referred to as moondogs.

supercell [99]**:** a vast, rotating convective system that can persist for long periods of time and bring damaging winds, hail, thunder, lightning and tornadoes.

supercooled liquid water [44]**:** liquid droplets in the atmosphere that have formed at temperatures below 0°C.

supersaturation [42]**:** air with a relative humidity over 100% is said to be supersaturated.

surface temperature [55]**:** the temperature of the air in the lowest few metres of the atmosphere.

synoptic observations [15]**:** a set of observations that are reported frequently (usually hourly) throughout the day and night.

T

teleconnections [92]**:** remote impacts of the weather conditions in one location on weather conditions in a different part of the world.

temperature [36]**:** a measure of how hot or cold an object (or mass of air) is. On a molecular level, it is related to the speed at which the molecules move.

temperature inversion [29, 38, 58]**:** a location in the atmosphere where temperature increases with height.

temperature proxy [128]**:** something that gives an indication of past temperature, such as tree rings and ice cores.

thermal [57]**:** a rising plume of air that is warmer than its surroundings.

thermal (or terrestrial) radiation [48]**:**

infrared radiation that has been emitted by the Earth or the atmosphere.

thermistor, thermocouple [24, 25]**:** types of small electronic sensors that are used to measure temperature.

thermodynamics [8]**:** the transformation of heat into different types of energy.

thermometer [6]**:** a term for any device that measures temperature.

thermosphere [39]**:** the low-density outer layer of the atmosphere where temperature increases with height.

tipping-bucket rain gauge [20]**:** a type of rain gauge that collects rainfall in a container mounted on a pivot. When the rainfall accumulation exceeds a certain threshold, the container tips over and the tip is recorded.

TIROS-1 [11]**:** the first successful weather satellite, launched by NASA in 1960.

tornado [100]**:** a rapidly rotating column of air that extends from the bottom of a severe convective storm system.

Tornado Alley [100]**:** a stretch of the central and eastern USA where supercells and tornadoes tend to form.

total solar irradiance [49]**:** the total amount of incident energy from the Sun per unit area at the top of the atmosphere.

trade winds [68]**:** the easterly, Equatorward winds associated with the Hadley cells in the tropics.

triple point [76]**:** in a mature mid-latitude depression, this is the point where the warm front, cold front and occluded front meet.

tropical continental air mass [75]**:** a mass of air that originated over land in the tropics.

tropical cyclone [88]**:** a tropical depression whose winds exceed 33 m s^{-1}. In different parts of the world, they are known as hurricanes or typhoons.

tropical depression [88]**:** a low-pressure system that forms in the tropics, normally consisting of a mass of convective activity.

tropical maritime air mass [75]**:** a mass of air that originated over the ocean in the tropics.

tropical storm [88]**:** a tropical depression whose winds exceed 16 m s^{-1}.

tropics [84]**:** the part of the Earth where the weather is governed by the Hadley circulation, typically extending about 30° either side of the Equator.

tropopause [38]**:** the temperature minimum at the top of the troposphere.

troposphere [25, 37]**:** the layer of the Earth's atmosphere nearest the surface, marked by decreasing temperature with height and containing most of the weather we experience.

trough [74]**:** a long extension of low pressure.

turbulent flow [38]**:** in a fluid, turbulent flow is characterised by disorganised motion in eddies and vortices.

typhoon [88]**:** a name for tropical cyclones that form in the western Pacific.

U

ultraviolet radiation [47]**:** electromagnetic radiation with wavelengths between about 10 nm and 400 mm; a component of solar radiation.

Universal Time Co-ordinated (UTC) [14]**:** the time convention used by meteorologists all over the world. UTC is fixed to Greenwich Mean Time.

unstable [117]**:** in a weather forecast model, unstable activity causes the model to fail and produce unrealistic results. In the atmosphere, unstable conditions lead to rapid cloud development.

updraught [96]**:** the ascending column of air at the heart of a convective storm system.

upper air [25]**:** a term for the mid-to-high troposphere.

urban heat island effect [110]**:** the enhanced temperatures found in an urban area that occur both during the day and at night.

V

validation [132]**:** the verification of the performance of a weather or climate model by running it over a period where the weather conditions are known and checking its ability to reproduce these conditions.

vapour pressure [42]**:** the contribution to atmospheric pressure that is due to the molecules of water vapour.

veering [18]**:** the wind is said to be veering if it is changing in a clockwise direction (for example, from southerly to westerly).

visibility [21]**:** the furthest distance at which objects can still clearly be seen.

W

Walker circulation [91]**:** the east–west circulation of air over the tropical Pacific which varics with phase of the Southern Oscillation.

wall cloud [101]**:** the mass of cloud that encircles the updraught in a supercell and indicates a region where tornadoes are most likely to form.

warm conveyor belt [81]**:** one of the three air streams that maintains a mid-latitude depression, consisting of a stream of warm air that rides up the warm front.

warm-core anticyclone [75]**:** a type of anticyclone containing a deep column of descending air in its centre.

warm front [76]**:** a boundary where a mass of warm air rides up over a mass of colder air, leading to ascent and cloud development.

warm sector [80]**:** the area between the warm and cold fronts in a mid-latitude depression that contains warm, moist air.

water vapour [34]**:** water in the atmosphere in gaseous form.

waterspout [101]**:** a tornado over the sea.

weather lore [5]**:** a set of sayings, based on natural indicators, traditionally used to give rudimentary weather forecasts.

well-mixed gases [34]**:** most gases found in the atmosphere are said to be well-mixed as they exist in the same percentages in the lowest 100 km.

wet-bulb thermometer [16]**:** a thermometer whose bulb is kept moist and is used in combination with a dry-bulb thermometer to measure humidity.

wind shear [76]**:** a change of wind speed and direction with height.

World Meteorological Organisation (WMO) [14]**:** a branch of the United Nations that enables worldwide co-operation in weather observation and forecasting.

Y

yellow dwarf star [47]**:** an astronomical category of star that includes the Sun.